Computer Graphics and Animation for Corporate Video

by Stephen Wershing
and Paul Singer

Knowledge Industry Publications, Inc.
White Plains, NY

Video Bookshelf

Computer Graphics and Animation for Corporate Video

Library of Congress Cataloging-in-Publication Data

Wershing, Stephen.

Computer graphics and animation for corporate video / Stephen Wershing, Paul Singer.

    p. cm. — (Video bookshelf)
Bibliography: p.
Includes index.
ISBN 0-86729-239-3
1. Computer graphics. 2. Computer animation. 3. Video tape advertising. I. Singer, Paul (Paul R.) II. Title. III. Series.
T385.W474 1988
006.6—dc 19

                                              88-25938
                                                  CIP

Printed in the United States of America

Copyright © 1988 by Knowledge Industry Publications, Inc.,
701 Westchester Ave., White Plains, NY 10604

10 9 8 7 6 5 4 3 2 1

Table of Contents

List of Figures

Introduction

In *Computer Graphics and Animation for Corporate Video* we will be reviewing a new discipline within video. Why should you, as a corporate video producer, be interested in computer graphics? By using the same basic technology that has been available in video for the last five years, there is now equipment and software that offers new ways of creating images for recording on videotape. As the cost of equipment for producing this imagery has decreased, and as the need for video communications has been recognized and better funded over the same period of time, this new means of generating video has come within reach of the producer of private television. You, as one of those producers, are in a position to utilize these new techniques to enhance your programming at a time when the demands on the quality and utility of your product are coming under ever increasing scrutiny.

Computer graphics offer a means of correcting flaws in existing video and a means of visually communicating that which cannot be captured by a camera. Computer graphics systems enable the producer to create images directly from a conceptual stage rather than setting up scenes or manipulating existing video. Most important, computer graphics are production tools that make possible imagery that can capture and hold the attention of the viewer like no other production technique currently available. Capturing the audience's attention means a more memorable program, which translates into more persuasive and instructional potential for your product. The viewer's increased interest means a more convincing and enduring message. And the point of corporate communications, of public relations, of advertising and of training videos is to change the thoughts and actions of the viewer for as long as possible.

Current technology has caused us to expect a high level of sophistication in the realm of visual communication. The popularity of broadcast television is no small contributor to this mind-set. In the competition for the viewer's attention, advertisers spend extravagant amounts of money to deliver their messages by producing commercials and buying media time. In addition, some of the most prestigious computer animation houses in the world receive top dollar not only to produce commercial messages with their equipment, but to develop new technologies and styles for communicating them. Such firms find customers not only in companies that want tried and true production techniques and commercial approaches, but companies that aspire to rise above the milieu of broadcast television to deliver a message that stands out in the consumer's mind.

The instructional producer, therefore, is charged with the responsibility not only of presenting the appropriate material in an understandable fashion, but also with reaching the audience and communicating with it. Judith and Douglas Brush pointed out in their third report on private television, *Into the Eighties:*

> In today's information society... it is the *receiver* who has control
> of the communications process both in terms of his or her demands
> for information, the form in which they want it and whether or not
> they are willing to accept what you send them.*

Bombarded every night with the most sophisticated imagery available, then, the target of corporate communications is a very demanding audience. Its set of expectations increasingly includes the type of visual presentation most effectively accomplished with computer graphics.

Realizing that some form of computer graphics is becoming commonly expected of the media production facility, we turn to the problems of the unique demands of this new and different technology. In addition to an eye for composition, the hand of a painter is now necessary. Where the ability to conceive from scene to scene was once sufficient, now it is often useful to be able to think frame to frame. While up until recently, editors and videographers could pick up the advances in private television, the corporate communications professional is now expected to successfully interface computer graphics technology with his current production portfolio. The solution most frequently and successfully utilized is to accept a new specialist onto the production team.

During the development of computer-based editing and digital effects, the videotape editor became more and more specialized and grew more distant from the videographer. Now, as we enter the era of widely available computer-generated imagery, new professionals are called for in the video realm—the computer graphics artist and the animator.

As technology develops, it becomes easier for anyone to understand the computers that drive graphics, editing and special effects equipment. In the past, when a freelance producer first entered the business or when a corporation first took production in-house, one person did it all. As facilities grew and as the machines became more

---

*Judith M. and Douglas P. Brush, *Private Television Communications: Into the Eighties* (Berkeley Heights, NJ: International Television Association, 1981).

sophisticated, it became clear that the more specialized production personnel were, the more effective they were at extracting the full power of their equipment. The editor who spends all day at the ADO (Ampex Digital Optics digital effects unit) better understands how to take it to its limits in creating 3-D graphics than does the videographer who shoots one day and edits the next. Similarly, the artist who spends all day in front of the graphics system becomes more adept at exploiting its capabilities.

THE HISTORY AND DEVELOPMENT
OF COMPUTER GRAPHICS FOR VIDEO

The history of the application of computers to video is short but full. The speed with which new developments and improvements are being made is astounding. Television audiences saw the first computer-manipulated video effects less than 20 years ago. The systems that have become available in the past several years are capable of generating three-dimensional models, viewed in perspective, with the ability to shade and light the resulting objects to create realistic scenes. The degree with which these machines can render realistic images is a function of their computing power. With the miniaturization of memory and the advancement of microprocessors, however, this power can now be contained in a desktop computer.

With current technology, entire scenes can be created in a computer; many are so realistic that audiences sometimes have difficulty differentiating between the genuine and the computer-generated. Graphics systems are now capable of communicating with other computers to extract information from them for video presentation. An increasing number of systems are available with the ability to transfer design information from a CAD (computer-aided design) system into its own database. With these systems, models of proposed products can be transferred and displayed electronically. Even more exciting, Pixar produces a graphics computer (the Pixar Image Computer) that can interface with medical diagnostic imaging equipment, for example, to generate three-dimensional images of a patient's bones. These images can then be used in presentations to other medical professionals or in instructional programs for patients.

The Developing Acceptance of Computer Graphics Images

In *The Algorithmic Image,* Robert Rivlin compares the development of computer graphics to Thomas Aquinas' description of the stages of human perception of God: wholeness, radiance and harmony. In the early days of computer graphics, people who witnessed the work of the few early research centers stood in awe of what they saw (wholeness). As the technology became better understood, many different disciplines developed using computers for graphics generation: CAD/CAM (computer-aided design/computer-aided manufacturing), business graphics and computer composition of print materials. The central technology of computer graphics branched into many different fields (radiance). Finally, Rivlin hypothesizes, computer graphics are developing into a social force by coalescing and becoming part of everyone's life (harmony).* Shoppers touching the image of the product they want on a computer screen in a supermarket, video illustrators creating charts and graphs on their digitizing pads, and executives getting printouts of their schedules, which include clocks to represent the time of each appointment—these individuals do not consider themselves part of the computer graphics industry, but computer graphics have become an integral part of their lives.

For computer graphics to achieve this stage of acceptance, the value of the images must remain high. Simple graphics, which communicate thoughts clearly, and sophisticated graphics, which make complex concepts understandable, must be the norm. Producing graphics that are valued not only because they are attractive but because they *communicate* will be the basis for achieving this harmony.

THE FUTURE OF COMPUTER GRAPHICS

What can we expect in the near future? An apparent goal of engineers is to increase the speed and memory of computers. Each successive version of the graphics computer is capable of generating more sophisticated images and can produce them faster. We may see new ways of storing images that allow more to be stored in one device or may allow more complex images to be created. Computers with enormously increased internal working memory are likely. The use of videodiscs may become more widespread in conjunction with graphics systems.

One technique under investigation for increasing working storage is the RAM (Random Access Memory) disk. A RAM disk is computer memory that identifies itself to the central processor as a disk drive. It is a virtual device, meaning it exists as a device only in the computer's memory. When the computer uses the disk, however, it looks, acts, and answers to the name of a peripheral device. This is useful because computers can be equipped with memory far in excess of what the control chip can address. Some applications demand the use of more memory than can be addressed at once. When we use such an application, the computer compensates by breaking the program into manageable pieces and runs the program one piece at a time. This is one reason why, when we use a large program in a personal computer, we notice the disk drive light going on and off several times while the program is running. The drawback to this technique is loss of speed—it takes the computer time to load and unload peripheral devices.

RAM disks, then, allow the computer to work with more memory than the controller chip can address, but because the additional storage is entirely in memory, it can accomplish this additional capacity without the loss of speed inherent in employing a peripheral device. Entire images can be stored in the computer without drastically increasing the complexity or cost of the machine. Some machines that operate on this principle and other means of storage with high-speed recall already

---

*Robert Rivlin, *The Algorithmic Image: Graphic Visions of the Computer Age* (Redmond, WA: Microsoft Press, 1986).

exist in computer graphics. The Quantel Harry is one such device that uses hard disk storage for computer images.

Another anticipated advance, which can be extrapolated by examining the history of computers, is decreasing cost and size. Only a few years ago, it was impractical to attempt to produce graphics on anything less than a mainframe computer. Today, we have three-dimensional modeling packages available for use with microcomputers. What could once only be produced on a computer costing over $150,000 can now be accomplished on a machine costing $30,000. As prices and dimensions continue to fall, it may someday be possible to render such complex graphics as ray-tracing algorithms on a personal computer that costs only a few thousand dollars.

We expect that in upcoming years it will become easier than ever to create and manipulate images with computers. Most systems still require some degree of computer literacy, although we must compliment many engineers on creating very friendly systems. There are systems that allow the artist to form a palette and choose and mix colors as if the artist were actually using paint and brush. The Macintosh computer by Apple uses a system of icons to denote system functions as well as paint and draw functions, making it easy to learn to use the operating system. More progress in this area can be anticipated.

One issue also of concern to many video artists is resolution. Although images must be converted to NTSC (National Television Standards Committee) standard for broadcast or recording on tape, a higher resolution when creating an image results in a clearer finished product. Creating images in more than 525 lines and converting accomplishes some anti-aliasing and shading. For hard-copy output, in presentation graphics for example, higher resolution than video standard is imperative. The higher the resolution, the sharper the image becomes. With the dawn of high-definition television, enhanced resolution is called for. A standard of over 1000 lines resolution may not be so farfetched in the near future.

Most important, we anticipate the information content of graphic images will increase. Most of the uses for computer graphics we have discussed are effective because of their communicative or instructional content. To justify the expense and production time of sophisticated graphics, there must be more information communicated to the audience. We have stressed that computer graphics should not be used simply because the technology is available, but because the content will be enhanced by their use. If graphics and animation are to survive along with traditional video recording, they must be utilized as a more efficient form of transmitting information, because pretty pictures are evanescent and will be forgotten when the public's tastes change.

CONCLUSION

This book is intended to help the producer of private television understand the needs and capabilities of the computer graphics artist and to smooth the artist's incorporation into the production process. In it, we present the basics of computer-generated imagery, give a brief history of the growth of computer graphics, and present some ideas on how to effectively plan for and execute computer graphics segments in your productions. We do not foster any intentions of making anyone an expert in this field. We hope, instead, to accelerate your orientation to this new technology and to make your first applications of these techniques effective by learning from other producers' experiences.

We will present a fair amount of technical material in the course of describing the foundations and construction of computer imagery. Delivery of effective results in video and computer graphics relies on a thorough understanding of the tools used to create that product. Understanding the capabilities of available systems defines in large part how efficiently you can translate concepts into programs. Will the creation of a specific sequence or effect require an animation system, or can it be accomplished on a paint system? By selecting the appropriate tools you can achieve the most effective result at the lowest cost. And, as the technology of computer imagery becomes less expensive and more widespread, you may have the opportunity to incorporate more and more sophisticated processes into your own productions.

A technical understanding is critical when evaluating an outside rental facility or when choosing equipment for purchase. When a contractor or vendor says that his equipment is capable of rendering a real-time test sequence for client approval in wire frame or by reducing the number of control points defining the image, can you use this as the basis of an informed decision without understanding the meaning of the terminology? Even more critical, the ability to translate manufacturer specifications and equipment capabilities into benefits to the video department forms the basis for budget decisions. Preparing the department budget, describing the need to corporate management, or offering your department's facilities to other corporations is impossible without some understanding of the technical requirements and attributes of the equipment and processes involved.

In approaching the material in this text, we make the assumption that the reader knows very little about computer graphics and how they might fit into the video production process. For those readers who have been producing or working with graphics for some time, we ask your patience. We are very excited about the possibilities of computer imagery in private television and have undertaken this project in the hopes of including corporate video personnel who have had little exposure to it. This text, in fact, was designed more for corporate communicators who have not had the opportunity to make use of digital imagery in video than for members of large and technically advanced video departments. We make some assumptions regarding the reader's familiarity with video production technique, but feel it would be a mistake to assume a similar level of knowledge of computer technology and terminology.

1 The History of Computer Graphics in Video

A brief discussion of the history of computer graphics is required in order to develop an understanding of their impact on all video production and to project how corporate communications may be affected by computer graphics in the next few years. Unfortunately, compiling a history of such a convoluted network of interdependent but separate disciplines is difficult. It would be different if computer graphics were one field of art or science, with developments being made in a linear, if not uniform or predictable, fashion. Our problem begins in defining computer graphics as a field of endeavor. It could arguably be said that the creation of computer images is a unique discipline within art or communicative design. (That statement, of course, would still provoke heated discussions between artists and computer engineers.) In looking back, however, we find that computer graphics arrived from many parents. Important contributions have been made by computer engineers, software engineers, mathematicians, television directors, video equipment manufacturers, photographers, cartoon animators and artists who have never before worked in film or video.

What will we include when we say computer graphics? Clearly, all non-text output from a computer cannot be included. That would include paper graphs, presentation slides and computer screen icons, little pictorial representations of computer functions popular on user-oriented microcomputers like the Apple Macintosh. In addition, text also played an important role in the development of computer graphics. Character generators were some of the first computer graphics generators available to the corporate producer. But at the same time, we cannot exclude everything that we do not consider to be video computer graphics. Paint systems and animation computers were not individual creations, which suddenly made possible graphics for video in the same way that the invention of the electronic camera and monitor made television possible. We will attempt to limit our focus to those developments and applications within computer science and video production that most directly influence the world of corporate television.

1

This limitation, though, will still result in something less than a direct path to television graphics and animation. Part of the reason is that the pertinent inventions and technologies did not arrive sequentially from one or two laboratories, but from various sources, including technical facilities, highly respected academic institutions, and entrepreneurs' garages and basements. In fact, new techniques are still being developed. Ordering the events and accomplishments is difficult when we are still in the thick of innovation. Regardless, the following should give you a flavor of some of the events that brought us to this stage.

PICTURES FROM COMPUTERS

The earliest computers, essentially enormous calculators, performed mathematical functions and presented solutions to problems in numerical form. For many applications, complex calculations resulted in a discreet number of figures that could be easily understood and acted upon. For example, weather, curvature of the earth, and position of a target were some of the inputs for a missile guidance formula. The computer output would provide the artillery man with the correct trajectory so that he could adjust his cannon before firing. Unfortunately, most scientific applications did not lend themselves to such simple solutions. Researchers would manipulate numbers in search of patterns. They expected and received vast arrays of numbers on reams of paper. It was not long after computers came into being, then, that alternate forms of output were considered. Computers were programmed to output voluminous information in the form of charts, tables and plots. Solutions were presented in graphical form on paper, using teletype like printers and digital plotters, devices that drew lines on paper by moving pens that were attached to mechanical arms.

EDSAC

Graphic images created as a mosaic of numbers and symbols began decorating computer rooms in the 1960s, but computer graphics actually go further back than this. Probably the world's first stored program machine was EDSAC, in use at the University of Cambridge in 1949. On its face were a number of cathode ray tubes (CRTs), which were designed for diagnostic purposes. When EDSAC needed repairs, the screens would display some of the electronic workings of the machine. Programmers, of course, found other uses for the screens. David Wheeler of Cambridge remembers an animation of a Highland dancer, produced around 1950.

The Cathode Ray Tube (CRT)

One of the first breakthroughs in the display of graphics came around 1955, a decade after the invention of the computer; technical data output from the computer was displayed on a cathode ray tube (CRT) instead of on paper. Until that time, printers and plotters were the only options for output, and it had not occurred to anyone to adapt anything else for that purpose. MIT's TX-0 computer was outfitted with a tube similar to the scope of a radar, and graphic information was photographically recorded for the programmer. This was a major modification of the computing process. Instead of the programmer reading through hundreds of pages of output from a single problem, the same computations could be summarized on graphs and charts and could be saved as a set of slides.

Once graphics were established as an output method, engineers started using graphics interactively, utilizing the information on the graphics terminals for input as well. One such application was the SAGE air defense system which mapped trajectory information of incoming missiles, superimposed the paths onto a map projected on a screen, and allowed the technicians to select targets shown on the screen by using light pens. The computer determined and transmitted interceptor launch information, tracked the targets, and could alert the operator to any untoward actions of potential targets that the operator had identified for close attention.

SKETCHPAD

The next development regarding graphics came in 1962, when Ivan Sutherland wrote his MIT dissertation, ''Sketchpad: A Man-Machine Graphics Communication System.'' *Sketchpad* was a software system created on the TX-2 and Whirlwind computers at MIT's Lincoln Laboratory. *Sketchpad* accepted input from the artist in the form of free-form shapes made on a CRT screen with a light pen or reproduced predefined shapes on the screen using the light pen and a series of buttons. The artist pushed buttons to select the desired shape or function: draw a line between two designated endpoints, draw a circle based upon a specified center and radius, rescale or copy items within a specified window, and so on. *Sketchpad* was the first system to allow input to a graphics program with an almost immediate effect on the screen, it was the creation of real-time interactivity.

One of the major breakthroughs of Sutherland's work was that it accomplished the separation of model from artifact, a fundamental for today's computer graphics uses and capabilities. The user of the system recalled or created an object that existed as a model within the computer. The output was merely a representation of the model, whereas in the graphics prior to *Sketchpad,* the graphic was the object. The artist was no longer simply the recipient of graphic output, but could directly interact with the model being created. The function of the output was to close the feedback loop between the model and artist, and not simply to reveal the answer to a question.

This development was also the first to implement the concept of ''user friend-liness.'' Artists and engineers could create defined or free-form designs with the light pen and function buttons, eliminating the need to use complex lines of code. Systems engineers were beginning to recognize the enormity of the programming challenge required to build a graphics system that did not require the operator to have substantial computer programming ability; engineers and illustrators were interested in (and employed to) generate designs and had little opportunity or desire to spend considerable time learning to program a computer.

The remarkable idea behind *Sketchpad* was an invention that later became known as the DPU (display processor unit). The DPU converted digital information output from the CPU (central processor unit) of the computer into analog information, which could be displayed on a CRT or plotter. The operator describes to the CPU that he wishes, for example, to draw a line and then inputs the end points via light pen. The central processor calculates all points between the ends and feeds the locations of points along the line to the DPU. The DPU then directs the electron beam in the CRT along those coordinates, and the result is a line on the screen.

REAL-TIME PICTURES

Early graphics systems produced images by sequentially stimulating phosphors on the screen one point at a time. Since the images were meant to be recorded by a photograph, this posed no real problem. The camera was trained on the CRT, and the shutter was left open long enough for the computer to plot all the points defining the image. Unfortunately, as a means of real-time or interactive output this was not very practical. A phenomenon known as persistence of vision allowed the viewer to perceive a whole image after seeing a rapid succession of points of light. It was impossible, however, to refer back to a finished graphic because the points of light were not continually renewed, but activated momentarily and left to decay. The production of an image that could be viewed for a length of time required the computer to activate the dots continually, stimulating the first dot as soon as it completed the last dot of the picture and running through the sequence fifty or so times per second. This constant transmission of data to ''refresh'' the image was a significant demand on the computers of the time.

The first step toward a solution was to use the CRT's electron beam to paint swaths of light on the screen rather than calculate a large number of independent points. A simple comparison would be to use a pen and ruler to draw a line on paper instead of a typewriter's underline key. Using the typewriter, you can only place one character width on the page at a time and must repeatedly hit the underline key to create a line. Using a pen and ruler, however, you can draw an entire line in one motion. If information were expressed in terms of lines and curves rather than points, the amount of information the computer needed to refresh on the screen would be substantially reduced. Instead of using a multitude of points, an image was created when the computer sent a discreet number of lines and curves to the CRT. Since the picture was composed of illuminated lines drawn sequentially on the screen, these compositions became known as vector refresh displays (see Figure 1.1). They were essentially converted radar displays and were in service before much attention was drawn to computer graphics. SAGE, *Sketchpad* and other pioneering systems used this vector refresh technology.

DEVELOPMENT OF THE DVST

The vector refresh displays, which made possible early developments in computer graphics, were not compatible with television tubes or circuits. The method of creating a picture on these modified radar scopes proved to be a major limitation. The vector display plots each line segment in the image sequentially. Since the phosphors that glow to form the image decay rapidly, the computer had to redraw the picture

Figure 1.1: Vector Display Tube.

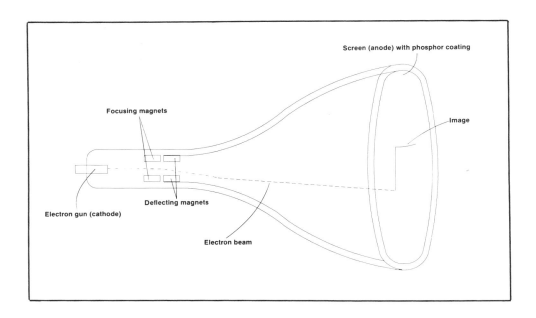

Figure 1.2: Direct View Storage Tube (DVST).

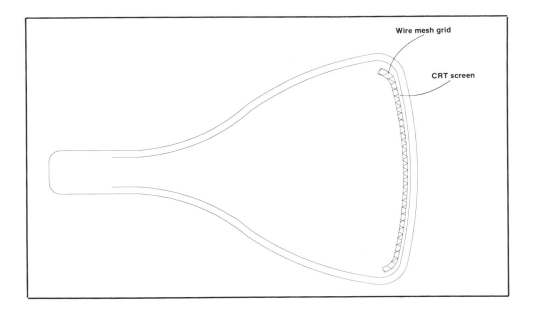

continually. When a complex image composed of many line segments was drawn, this continual redrawing was slowed down, and the image appeared to flicker.

Another drawback to these systems was their inability to reproduce different intensities or color. A phosphor was either stimulated or it was dark, no intervening shades of grey were possible. This attribute mattered little to the draftsman. The line of light from a DPU was not different from a line of ink on paper. However, the capabilities of this system held no real appeal for the illustrator. Compounding the problem was the price. The systems of the day (mid-1960s) cost $150,000.

The DVST (direct view storage tube) was introduced in 1968. Its lower cost brought computer graphics to the masses. A person no longer had to dedicate over $100,000 and six months or more to assemble hardware and debug software to create computer graphics. This invention was a CRT whose electron beam did not directly stimulate the phosphors on the inside of the screen. Between the electron gun and the screen was a wire mesh. (See Figure 1.2.) Points on the mesh that were struck with electrons from the cathode would discharge electrons onto the phosphors coating the screen. One hit on the mesh would cause a stimulation of the corresponding phosphor for as long as the mesh received an electric current. This meant that a graphic could be displayed on the screen for any length of time without having to be continually refreshed by the computer.

The DVST had its share of shortcomings. For example, lines could not be selectively erased, and it was difficult to view in a brightly lit room. However, the fact that it was cheap overshadowed its deficits. Also, the substantial computer power required to continually refresh the vector display was no longer required. Engineering workstations utilizing the new technology soon flooded the market, led by the ART system from Computer Designs, available at a modest $15,000.

RASTER GRAPHICS

The DVST and vector refresh display were still not capable of producing the type of graphics we know today. In a vector graphic, all images are created by a series of lines. The newer systems allow colors, but the nature of the vector graphic rules out solid surfaces. The simulation of a solid surface can be created by a series of lines drawn close to each other (see Figure 1.3). The breakthrough that resolved these remaining problems also came in the late 1960s with the advent of the frame-store driven raster display. "Raster" refers to the way a picture is displayed. Rather than illuminating only those points that define a graphic, a raster display has the electron beam visit each point on the screen, one at a time and in order from top left to bottom right. As the beam touches each point it varies the brightness and color as necessary to create the desired image on the screen. (See Figures 1.4 and 1.5.) You may recognize this as the method by which a video picture is displayed on a television tube. With the advent of the raster, computers could output information in terms a television would understand.

Figure 1.3: Vector Graphic. (See Plate 1.)

Courtesy Wavefront Technologies

The Pixel, or Picture Element

The fundamental building block of the raster image is the picture element, or pixel. The image on the screen is broken down into a grid of individual points, each of which is a pixel. To the frame buffer, each of these pixels is an individually addressable memory location, and each pixel has a designated location on the screen. Each memory location in the frame buffer contains information on how the pixel will appear when placed in its assigned location on the screen.

The amount of information contained in each pixel is called its depth and is expressed in bits, the fundamental building block of computer information, which can assume one of two values, 0 or 1. A pixel with a depth of one bit can be either on or off. It cannot assume any value designating brightness or color, it has only the two most extreme states from which to choose. Each additional bit assigned to an individual pixel increases exponentially the amount of information it can contain. A one-bit pixel can assume a value of 0 or 1. Two bits allow a pixel to assume four values, 00, 01, 10 and 11, each bit taking its component values 0 and 1 in turn. A pixel that is three bits deep can assume one of eight values: 000, 001, 010, 011, 100, 101, 110, and 111.

Figure 1.4: Raster Scan Format.

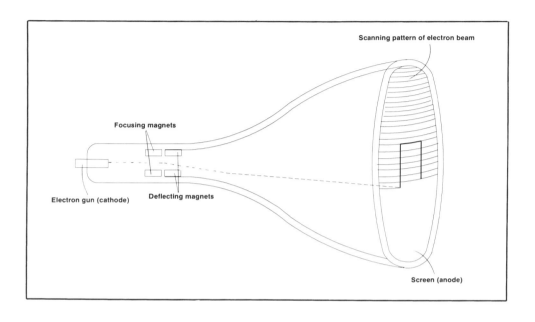

Figure 1.5: Raster Graphic. Car on Painted Background.

Courtesy Wavefront Technologies

In black-and-white systems, this pixel information controls the brightness of each point on the screen. In a color CRT, the computer can use the information contained within the pixel to control each of the three electron guns in the CRT (one each for red, green and blue) independently to vary a pixel's brightness and color. By varying the intensity of each gun, many color variations are possible. Systems are widely available today that define each pixel with 32 bits, allowing 16.7 million color possibilities per pixel.

The Frame Store

The frame store (or the similar frame buffer) is the same type of computer memory that forms the basis of today's digital frame store or still store device. The medium on which the video information is stored may differ in today's fancier models, but the function is the same: convert a screenful of information into a stream of digital values, one for each point on the screen. As the electron beam moves methodically down the screen, each of the corresponding values in memory is transferred to that point on the screen. Thus, the workspace within the computer is free to allow the operator to change or replace the image. Whenever the operator is ready to send a particular image to the screen, the computer replaces the existing matrix of values in the frame store with new values that together compose the altered or new image. In many systems this function is performed automatically and continuously, giving us interactive control of the image on the screen. Whenever the operator changes the image in the computer, it is instantly updated in the frame store and, therefore, displayed on the video screen. Regardless of what the computer is involved in doing, the picture on the screen is continually refreshed from the frame store, eliminating loads of computing to renew each point or line on the screen. This released the computer from overwhelming repetitious computation and simultaneously relieved the viewer of the decay and flicker problems.

The Visual Display Unit

The first of these raster displays, called the visual display unit, could do little more than store crude monochrome characters in 80 columns and 24 lines. They would manipulate characters of 9 x 9 dots along the raster. The dots would be laid out in a pattern on this 9 x 9 grid to form letters, numbers, and so on. The computer would have a repertoire of, say, 96 characters from which to choose. These standard characters were described by a code called ASCII (American Standard Code for Information Interchange). (ASCII was originally designed to describe to the computer the location of a specific character on the type drum of a teletype-like printer, the most common output device for the early computers.)

As computers and their frame store units became more powerful and less expensive, the output no longer needed to be limited to a small number of crudely described characters. Information about individual dots on the screen, such as brightness, intensity and color could be stored. Rather than relying on a limited library of characters or composing pictures from pre-determined shapes, these points could be used individually to compose arbitrary pictures. It was now possible to turn the entire video screen into a free-form picture, with a resolution as fine as dots on a television screen.

EARLY ANIMATION

The first computer-animated film, which helped popularize the vector graphic display as a medium for scientific presentation, *Simulation of a Two-Giro Gravity Gradient Attitude Control System,* was produced in 1963 by E. Zajac. This was not computer animation as we know it today, but rather an animated film composed of computer graphics still frames. Stills from this film were published in the journal *New Scientist* in February 1966, helping to alert the scientific community to the possibilities of this new technique for displaying scientific principles. One reason this technology was accepted early-on was Zajac's revelation that the cost of filming 42 seconds of *Simulation* was (startling by today's standards) $30.

Frank Sindon, working at Bell Laboratories, developed the first computer animation specifically for an educational presentation. His instructional film, *Force, Mass and Motion,* showed Newton's laws of motion in operation and demonstrated the inverse square law. It was produced by photographing individual frames as the computer created them. Now that computers were producing instructional programming, one major technological problem remained before computers could be used to produce video. The operator needed a method to give graphics instructions to the computer.

ANIMATION LANGUAGES

BEFLIX

Programming the computer to produce animated graphics was a job only experienced programmers were able (or had sufficient interest) to do. To assist the non-engineer, several high-level languages were developed for the creation of graphics. The first of these was BEFLIX (short for "Bell Flicks"), a language written by Ken Knowlton in 1964 at Bell Labs. With it, the operator could manipulate a matrix of 252 x 184 values of three bits each, allowing the generation of eight gray levels. This was the language used by Knowlton and experimental filmmaker Stanley Van Der Beek to create the animated *Man and His World* display for Expo '67 in Montreal.

EXPLOR

The first improvement on this system was another language also developed by Knowlton in 1970, EXPLOR (Explicitly Provided 2-D Patterns, Local Neighborhood Operations and Randomness). Less cumbersome than its name implied, the system allowed the artist to manipulate a 240 x 320 grid, and each point could assume a value of 0 to 255, which would denote both color and intensity. This grid system was the first method of storing information in the raster-scan format. Thus, although of much lower resolution than the graphics systems before it, pictures could remain on the screen for any length of time, and motion could be achieved directly on the screen without an intermediate photographic recording. The system was designed chiefly for motifs and mosaics, but creative users moved beyond those boundaries.

GENESYS

Another system of the time brought computer graphics one step closer to traditional film animation. It was GENESYS, developed by Ronald Baecker in 1969 as his doctoral thesis for MIT. It was a picture-driven animation system—all information defining an image had to be in graphical form. The new and interesting aspect of this system was its representation of dynamic behavior. The animated objects within the picture were described to the computer and electronically superimposed over a background. This demonstrated that dynamic information (a description of the motion of the foreground object) could be abstracted, modeled and generated just like animated films before computers.

These languages and others developed in the 1960s were experimental or prototypes, and none gained commercial or popular usage. Early computer graphics researchers were excited about the prospect of animating images created by the computer because the processes lent a new dimension to the pictures they had created on the screen and because of the potential for commercial usage. Knowlton, in writing about his work in animation, pointed out that the speed of calculation and automatic progression of one image to the next, using a computer and automatic film recorder, made possible the production of films that would have been far too costly or difficult with traditional methods. Most of all, however, these early languages failed to anticipate the coming of raster-scan technology and faded from the scene along with the vector refresh systems that supported them.

COMPUTER GRAPHICS MEET TELEVISION

Computer graphics met video production formally through analog machines built as special effects generators. One of the first was Scanimate (manufactured by Computer Image Corporation), which could perform some of what we think of as video special effects. It could manipulate art cards to zoom in or out, rotate on any axis, stretch or squash, and so on. But the popularity of these systems was short-lived. The machines were expensive and had very limited capabilities, so producers and audiences tired of their realm of effects quickly. Computers had been introduced to the world of television, however, and producers had gotten a taste of what these number-crunching machines could do.

In 1976, television viewers began to see computer graphics used as a production technique beyond the manipulation of art shot with a camera. The first successful commercial made by computer animation was "Think Electric," produced by Electronic Arts, Ltd., of London. Input and output were made on paper in wire form (a skeleton of an object made with lines, but without a surface), which would later be colored either by hand or photographically. This system also made titles for the "New Avengers" television series and the 1980 World Cup soccer matches.

Weather Central

One of the next developments in computer graphics for television came in 1978 with the development of the broadcast weather system. Rather than create entirely new images, this system colored and enhanced information received from weather radar. The concept was developed in the United States by Terry Kelly, president of Weather Central, and Richard Daly.

Daly had worked as a student on applications of the MCIDAS (man/computer interactive data acquisition system) developed at the University of Wisconsin. Weather Central was a consulting firm serving television and radio stations and various aviation companies. Daly went to Kelly in 1979 with the idea of sending computer-generated weather maps directly to television stations for inclusion in their nightly weather reports. They worked together on the idea and installed what is considered the first broadcast weather system in the United States at WNEP in Wilkes Barre, PA, in 1980. This image enhancement continues to play an important role in science and television. Indeed, with paint systems being increasingly utilized to "correct" location footage, the boundary between image generation and enhancement is becoming blurred.

AVA

In 1981 Ampex introduced AVA. AVA was a two-dimensional paint system based on a design by Alvy Ray Smith of NYIT (New York Institute of Technology), who saw a need for a flexible method of designing visuals with no requirement for computer literacy. The process begun by Sutherland in the 1950s had come to television for transmission into everyone's living room. Unfortunately, as with the early vector refresh systems, it carried a very high price tag. AVA was originally marketed at $200,000, hardly a price that brought the system within reach of most design studios. However, as has been the case with technology is general, other less expensive systems with extended capabilities began to appear.

CONCLUSION

The memory of the expensive and limited first computer graphics systems for television, like Scanimate, created some resistance to the introduction of their digital offspring. Producers associated computer-generated images with a limited range of overused effects. Dolphin Productions, which mastered the analog effects early on, developed graphics for the Public Broadcasting System's presentation of "The Scarlet Letter." The computerized product deliberately had the look of hand animation. Alan Stanley, president of Dolphin, even claimed that "you would never dream it was computerized."

Once computers became popular for uses other than graphics in the office and home, characteristics of the ubiquitous computer monitor refreshed producers' dislike for computer imagery. Many people, some video producers among them, began to believe that the rough looking letters and numbers and the jagged lines they saw on their minicomputer monitors were all a computer graphics system was capable of generating. Some came to believe that anything else was not really computer graphics. One client of a major video production house insisted in 1983 that the "jaggies" be put back into the image of his product so as to appear computerized, thus wasting lots of expensive anti-aliasing software. (See Chapter 3 for a more in-depth discussion of anti-aliasing.) His commercial, like many less sophisticated ones before it, persuaded some advertising agencies that jagged-looking aliased lines were a fact of computer life. They, therefore, looked no further at the possibilities.

Eventually, Robert Abel and Associates, Digital Productions and Cranston/Csuri persuaded the advertising world that computer-generated images could appear realistic and dramatic, and could sell products. One example is the phenomenally successful series of commercials produced by Abel and Associates for Levi's. One of these spots, "Brand Name," produced in 1977, still holds the record for viewer recall of any television advertisement since 1933. That kind of product recognition caught a lot of attention among advertisers and their agencies. This, coupled with the wider availability and lower cost of computer-imaging equipment, made digital graphics a part of everyday television programming. Intros, IDs, commercials and newscasts are full of images generated by computer. The progress and propagation of computer systems available to any video producer is startling. Today, a wide range of systems with a commensurate range of capabilities is available, and numerous methods have been developed to assist in imaging many different types of objects. The rest of this book will describe some of these methods and systems and will provide some pointers on how you can incorporate them into your production repertoire.

2 Uses of Computer Graphics

As was made evident in Chapter 1, computer graphics originated as a discipline separate from media production, and it continues progressing as a field in which video production is only a part. It was many years before computer graphics collided with the realm of video production. Today, the applications of computerized imagery overlap broadcast television and other media considerably. Because some specific computer-generated image was not intended for inclusion in a video program or because the output of a specific applications package* is not routinely transferred to tape, however, does not make such uses inappropriate. In fact, many of the most common uses of computer graphics in video were developed for other purposes, such as missile guidance and flight simulation. In our effort to make this text most useful to the video producer, we will touch upon some applications of computer imagery that do not typically end up in video programs. Of course, the primary focus of this chapter will be the most frequent uses of computer graphics in nonbroadcast video.

In this chapter, we will address many functions of computer graphics within the realm of private television. None of the applications will be appropriate for all productions. Many of the uses we will describe would be inadequate reasons to purchase a graphics system; few could be used as sole justification. We point them out anyway because, once a system is available, the utilization of the computer to accomplish some of the described effects is more efficient than relying on the previous methods. Graphics generators, for example, can be used to replace art cards. Obviously, a production facility does not require a graphics computer to produce an image, which a few years ago would have been produced on an art card. Having the availability

---

* An applications package is a commercially produced computer program, which allows the user to accomplish the goal of computing without having to learn a programming language. Wordstar, Visicalc and Lotus 1-2-3 are applications software; BASIC and COBAL are languages.

of a character generator or paint system, however, makes production of the image faster and easier, and renders it in a form that is more readily altered to conform to the preferences of the client.

The general progression in this chapter will be from simple to complex and there will be no stratification of the applications by importance. Some of the applications, which sound simple here, may actually be very difficult to produce without a computer. Many of the uses of computers will involve a combination of applications.

GENERAL PRINCIPLES

One of the principal reasons for the popularity of the computer is its ability to produce powerful, attention-grabbing images. Broadcast television has certainly capitalized on this ability, and the desire for more powerful imagery to capture viewers' attention is one of the driving forces in advancement of the medium. One unfortunate result of this technology, however, has been the development of the demanding viewer. The discriminating eye of the demanding viewer often places the corporate producer in the position of creating video programs that must, to some degree, compete with the most sophisticated broadcast images. Even the casual television viewer comes to expect a certain pace and ''look'' to information transmitted via the video screen. This is not to suggest that only commercials and music videos can succeed in communicating, but it is imperative for the nonbroadcast producer to be sensitive to this bias in the viewer's mind.

Uses of Computer Graphics

The uses of computer graphics can be roughly divided into two categories: the addition of visual excitement for the reasons outlined above, and to portray that which cannot otherwise be observed by the naked eye. The latter category includes phenomena that cannot be readily viewed because of scale or access. While these two basic categories describe the function of the graphics, they in no way determine application. Either could be promotional, instructional, or entertainment and, with any luck, most uses include all three. Specific examples will be discussed throughout the chapter. Training, corporate communications and even commercials can be instructional. Most are promotional, whether the video is promoting a product, a technique or the company's point of view. And it must be entertaining, or any retentive value will be lost as soon as the screen goes dark.

ELECTRONIC ART CARDS

Producing images on a graphics computer is often much easier and faster than creating traditional art cards. Advanced graphics studios, in fact, are switching from more traditional methods of illustration and titling to using graphics computers capable of outputting hard copy and transparencies.

Figure 2.1: Two-dimensional bar chart. (See Plate 2.)

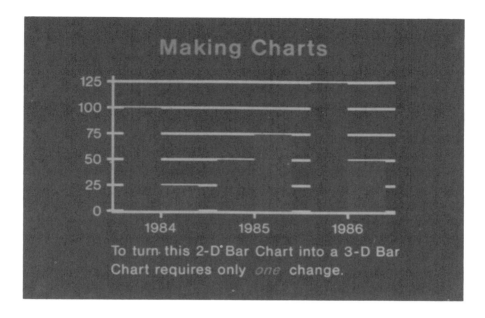

Courtesy Artronics Corporation

Figure 2.2: Three-dimensional bar chart. (See Plate 3.)

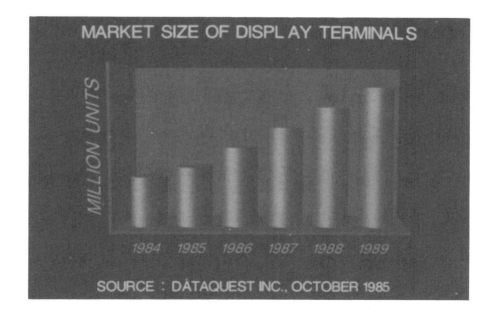

Courtesy Advanced Entertainment Associates

THE CREATION OF TITLES AND GRAPHS

The use of any but the simplest graphics generators provides considerably more flexibility in imagery than pen and pencil methods. The most obvious uses for computers, then, are the creation of titles and graphs for video. Letters, which can then be positioned, colored, and manipulated in visually interesting ways, in a number of fonts are available as standard in most systems. They can be used alone or as one visual element in a complex image, which can be created right on the screen.

Character Generators

The simplest computer graphics machines, character generators, are designed specifically for this type of task. The more sophisticated systems have this capability too, but character generators were built for this alone. Character generators are available now that will perform more complicated tasks. Some have internal paint functions, some can do color cycling (which will be discussed later in this chapter) and other animation techniques. We will not discriminate between character generators, paint systems, or animation workstations in this chapter. For our purposes, character generators are television typewriters. Most models available today are capable of more than this, and we consider them graphics machines.

Creating Graphs

Graphs can be produced as well as titles and charts. Graphs can be rendered in two or three dimensions, in many colors, and can be built before the viewer right

Figure 2.3: Composite Graphics. (See Plate 4.)

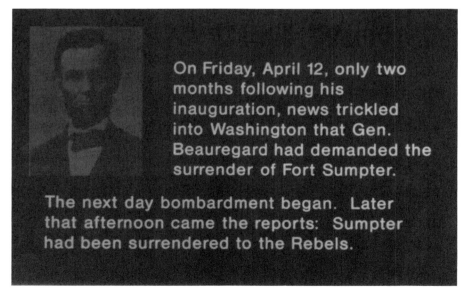

Courtesy Artronics Corporation

on the screen. (See Figures 2.1 and 2.2.) This build can be accomplished through editing or performed in real time, depending on the sophistication of the graphics system. In either case, the technique is easier than generating a series of art cards and capturing them individually to produce the same effect, and it is certainly less time-consuming.

Creating Backgrounds

Similarly, backgrounds can be created for titles or effects. (See Figure 2.3.) More than the standard black or color fields or live video are available. Ramps and color gradations can be used in place of a solid color background. Repetitive patterns can be produced in subdued shades to give the impression of texture without being distracting. The company logo can be rendered in a bas-relief to personalize the background. (See Figures 2.4 and 2.5.) Illustrations can even be produced as a background or as a frame for titles. Still photographs and frames of live video can be captured, the contrast reduced and the scene colored in a monochrome for use as a background.

Altering Images

Apart from originating images in the computer, paint systems can be used to alter or "clean up" images captured form traditional art cards. Many computer graphics systems allow images fed by a video camera to be "scanned" into the computer for manipulation as if it had been created on the system. It is not unusual to capture some video noise or reflections when scanning in an art card. This image can be altered by the paint system to appear more acceptable. Black levels can be lowered for more effective keying, borders can be introduced, colors accentuated, or slightly out-of-focus elements can be clarified.

Producing Storyboards

The drawing capabilities of the computer can be used to replace pen and pencil in the production of storyboards. These systems will require some form of graphic hard-copy output, but the quality need not be spectacular. Images can then be stored and changed without requiring the entire board to be redrawn.

In broadcast advertising, storyboards are drawn and approved on a preliminary basis. Entire commercials are then created in "mock-up" by drawing the scenes in full color on full-size cards and shooting them with the same camera moves that will be used at the final shooting. Recording these drawings in this fashion is called *animatics* and allows the client or creative director to view a close approximation of the finished commercial. Animatics, too, can be created with computer graphics rather than drawing the scenes with pen and ink and watercolor on art cards. If the finished piece will be animation, this offers the opportunity to present the actual key frames of the animation for approval. When the client or producer has seen the "rough" or simulation of the sequence, the viewing will have included actual graphics, which will be used in the production.

Figure 2.4: DuPont logo treatment. (See Plate 5.)

Courtesy Marshall Productions

Figure 2.5: Simulated metallic logo. (See Plate 6.)

Courtesy Wavefront Technologies

Logo, Intro and Segue Treatments

The need to attract attention and the limitations of budget make logos, openings and segues among the most popular applications for computer graphics. These pieces are brief, occupy opportune positions in the timing of programs and are reusable. Their repeated use in a production or series of shows helps make them cost effective. Developing a company identification segment for use at the beginning of all a firm's programs can demonstrate the impact of computer graphics to a management otherwise averse to the high price of the technique. Logo treatments can be a good introduction to computer graphics for a company considering more extensive use of this type of imagery or the outright purchase of a system. (See Figures 2.6, 2.7, and 2.8.)

VIEWING THE INVISIBLE—SIMULATION

Computer graphics can be very useful in portraying a host of events, which for reasons of scale or availability are not recordable by traditional photographic techniques. One of the most popular is the depiction of process or function. Consider the inner workings of an automobile engine. Attempting to explain the functions of individual parts could get pretty confusing without some illustration. Actually cutting an engine in half and pointing in a camera will do little good, however, because the motor will not function in pieces. Artistic representations and cutaway views of the cylinders and block, however, can be used to demonstrate the operation of a car engine through computer simulation of the engine.

Figure 2.6: Computergrafik Weltweit logo. (See Plate 7.)

Courtesy Artronics Corporation

Figure 2.7: New World Center logo. (See Plate 8.)

Courtesy Artronics Corporation

Figure 2.8: Electric Paintbrush logo. (See Plate 9.)

Courtesy Electric Paintbrush, Inc.

Teleconferencing provides a convenient example. The scale of earth stations communicating via satellite is too large to capture with a camera, and the signals are invisible. The process can be clearly illustrated, however, with a graphic depiction. An artist's rendition of dishes that are sized out of proportion to the earth, littered across a continent and bouncing little lightning bolts off satellites, demonstrates the flow of information around a teleconferencing network. Not very elegant, perhaps, but effective; and creative minds can come up with a host of alternatives. It is not difficult to recognize the value of this approach when attempting to describe systems where a conference may have video originating at several points and many receiving sites capable of feeding back audio. Imagine working for a firm that sells teleconferencing services. A customer arrives who has never had the occasion to even participate in a multiple-site conference. Can you imagine how well an animated segment would explain your firm's product?

A commercial produced by Robert Abel & Associates for STP oil treatment provides another example. The viewer takes a computer-generated ride through an engine along the route of the motor oil. The spot exemplifies many of the uses discussed so far, albeit not entirely for their instructional value. (See Figures 2.9, 2.10 and 2.11.) The animation makes a point, though, about the functions of the motor oil and STP, and creates an attention-getting and memorable production.

MARKETING APPLICATIONS

Product Styling

Marketing professionals have discovered that computer graphics can be cost-effective when considering product styling. It is very expensive to produce prototypes of products or packaging. Designs can, however, be fed to the imaging computer to enable the product to be visualized before it is constructed. The computer can be used to generate new labels for products or construct entirely new dispensers. (See Figures 2.12 and 2.13.) It can also place the product on a simulated kitchen table or grocery store counter so the manufacturer can evaluate the product's visual appeal in relation to its competition. Some consumer-goods manufacturers now do this as a matter of course. In this process, a video crew tapes a product on the supermarket shelf. The tape is brought into the post-production suite and fed to the computer. Using the graphics system, the current product is airbrushed out of the tape and replaced with the proposed packaging. The resulting imagery of the new package on the shelf can be used in consumer reaction tests or to promote the new product to supermarket chains before release. Any negative reactions to the package can lead to necessary design alterations before the prototypes are made.

DESIGN AND ENGINEERING APPLICATIONS

Automobile manufacturers are very enthusiastic about using computer-generated simulations because of the enormous expense of building prototype cars. Design data

Figure 2.9: A computer-generated ride through an engine. (See Plate 10.)

Courtesy Robert Abel & Associates

Figure 2.10: A blueprint car. (See Plate 11.)

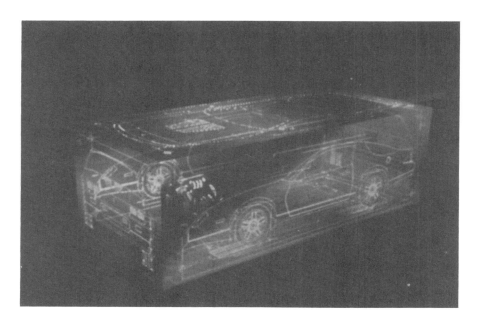

Courtesy Robert Abel & Associates

Figure 2.11: A computer simulation of a car.

Courtesy Robert Abel & Associates

can be fed to the graphics computer, and the car can be rendered. (See Figure 2.14.) Color and surface properties can also be input. At its most extravagant, the facsimile auto can be driven through different environments. Computer animation has created broadcast commercials for auto manufacturers, national brand lubricants, and even for a cellular phone company. Even if the images are produced in two dimensions on a paint system as stills for inclusion in a marketing video, there are tremendous dollar savings to be realized over constructing an actual model.

NON-TRADITIONAL SALES APPLICATIONS

The creation of models to evaluate product styling is not limited to traditional consumer products. Some production facilities have real estate development firms as clients that want representations of proposed buildings. Perhaps the most dramatic example of this application is Omnibus Computer Graphics' representation of Expo '86. The objective was to sell tickets to the event several months before it opened. The problem was that the fairgrounds had not yet been built. From architectural drawings, Omnibus created a three-dimensional fair and used it to produce a thirty-second broadcast commercial. The result was sales of ten million tickets over the two-month period before opening, prior even to breaking ground.

Real Estate Applications

This sales application has appeal for real estate developers that are now facing the twin specters of a soft commercial market and high construction costs. By creating computer-generated models of buildings, the developers can take their prospective

Figure 2.12: Morton salt containers constructed within the computer. (See Plate 12.)

Courtesy Cubicomp Corporation

Figure 2.13: Clorox bottles. Experimenting with new labels. (See Plate 13.)

Courtesy Cubicomp Corporation

Figure 2.14: Computer-generated simulation of a Camaro. (See Plate 14.)

Courtesy Robert Abel & Associates

tenants through office complexes without having to build the structures. (See Figure 2.15.) Prospective lessees do not seem to derive much of the feel of a building by looking at a scale model, so these walk-throughs can be very valuable. By offering facsimile tours before the structure is erected, it is possible to obtain leases and down payments before ground is broken, drastically reducing the costs of financing during construction and finishing. It can be the building's acid test as well; if tenants cannot be found for a certain percentage of the floor space before ground is broken, the project may be scrapped or scaled down, reducing the possibility of a costly failure.

Computers can render images of building interiors to evaluate interior design. Sequences can be generated to simulate the progression of the sun and the resulting shadows across the room. Finished offices can be approved before any materials or furnishings are purchased.

Computer animation has been used to assist developers in presenting their plans to local governments, and even to win legal battles. In one case, construction of a building was held up because some people in the neighborhood claimed that the proposed new structure would cast a shadow over their buildings and thus darken the whole area. Computer modeling of the area and the sun's progression across the area demonstrated that this was not true, and the resulting videotape provided the visual demonstration. Construction was allowed to resume.

Figure 2.15: Architectural rendering of a city block. (See Plate 15.)

Courtesy Cubicomp Corporation

ALTERING EXISTING VIDEO

Computer graphics and animated sequences need not be used independent of other production techniques. Computer imagery can be combined with images captured with the camera to produce finished images.

Computer-Generated Sets

Computers can provide sets for talent working in an empty studio. For very elaborate productions, this can save money and sets can often be produced more quickly than the stick-built version. When Chevrolet wanted to announce the redesign of the Corvette several years ago, they wanted to display it driving through a tunnel lit by fluorescent lights. Robert Abel demonstrated that they could save $50,000 by building the tunnel in the computer and driving the real car through the computer-generated image.

Computer-Generated Buildings

Mike Saz of MTI/Compugraph Designs in New York designed a building for an episode of the television series "Omni: Visions of the Future." He constructed several views of the exterior and interior hallways of the building, a medical laboratory of the future. The host appeared in the show standing in front of the building and walking through it. This approach allowed the creators of the program to give the building a futuristic look, with glowing walls and more perspective and depth than a real building would have. It also allowed scenes from several locations to be incorporated in the same imaginary facility.

When Kinetic Design Systems/Computer Productions, Inc. used a computer-generated haunted house instead of a location shoot, they were able to save enough money to hire a celebrity, Vincent Price, as spokesperson in a horror-movie style management film for Simon and Schuster Communications. The program, "Idea Power," dealt with fostering creative ideas and the entrepreneurial spirit within corporations and pointed out negative attitudes that can defeat them. Simon and Schuster wanted to use a horror-film style to tie all the negative pressures together. Charles Sleichter, executive producer of the program and president of KDS, realized that 15 minutes of edited programming with eight setups could not be accomplished within budget. Sleichter liked the concept of the horror-movie approach and thought ghosts would be an effective way to portray negative attitudes. He found, as well, that using computers to create the house, the ghosts and a number of other elements of the production enabled them to put everything together within budget.

RETOUCHING VIDEO IMAGES

Computers can also be used to retouch a video image. Sometimes, a graphic element can be created to cover part of the image, and it can be "pasted over live." That is, it can be left in place or moved with a DVE in post-production as previously taped action goes on behind it. Failing this approach, the video can be altered one frame at a time. Although a time-consuming, and thus expensive, process, it can be more economical than reshooting an entire scene. In some cases, reshooting may not even be possible.

Take, as an example, shooting a scene of a car driving down the street in daylight. Upon viewing the footage, you can see the crane and camera crew reflected in the windshield. Perhaps there was no way to avoid this, or maybe it was simply not feasible to reshoot. The computer's paint system can remove the reflections from the window, or even replace them with more appropriate images. This "invisible paintbox" technique can be used to remove shadows, trademarks, or even entire objects from live video. In one studio production, it was used to remove visible monofilament wires holding up a sign. In another production, the talent sat in a darkened studio under a single overhead light. In post-production, the effect was emphasized by using the paint system to create a "cone of light."

PRODUCT TESTING

Once the design information has been input, some of the high end systems will also accept engineering information about the model. Use, as an example, the previous discussion on automotive styling. Structural data about a car can be fed into the computer, and the operator can create simulations of the car crashing into walls at different speeds. Not only can information about the resulting damage be obtained, pictures of the probable damage can be viewed. Flaws in the design can thus be corrected before the prototype is ever built. The cost savings in this technique are obvious. The manufacturer not only saves a car, but the astronomical costs of building the tooling for a prototype are spared. This type of "what if" analysis may become as popular a use of graphics systems with engineering and design departments as the financial spreadsheet is for the personal computer in the accounting department.

TRAINING APPLICATIONS

Computer modeling also has many applications in the area of training by simulation. Manufacturers and users of large, expensive vehicles have found computer simulation to be a far less risky training method for operators than the real thing. The student is placed in the pilot's station of the vehicle and is surrounded by screens that simulate what he/she would see through the windows of the real vehicle. This technique has been used to train pilots for ships and test drivers for cars, but perhaps the most exciting application has been in the training of fighter pilots. Singer-Link Simulation of Binghamton, NY, is in the business of building simulation trainers for pilots. In fact, all airline pilots spend a large percentage of their training time in Link and comparable simulators. The practice is so widespread that federal regulations allow some pilots to move from flying one aircraft type to another solely on the basis of simulator training. Not only does this provide a safer and more comfortable environment for learning, but the cost savings are enormous. Singer has even taken computer-animated scenery and added it to the training video for promotion in its trade show exhibitions.

An exciting type of training being facilitated by computer is surgery. Along with observing actual procedures, interns can manipulate a computer-generated heart and observe the results of specific actions without working on a human. The pupil can attempt to compensate for his mistakes uninterrupted and observe the outcome of his decisions, rather than having the senior surgeon step in when an error is made.

TECHNICAL ILLUSTRATION

Computer graphics have been part of business for a few years. For example, computer graphics have become indispensable in financial analysis; vast arrays of numbers can be expressed in terms of a graph or chart. Computer graphics machines are now rapidly becoming more important in a number of other fields as well because of their ability to take complex arrays of data and render them in visual form, so underlying problems and trends can be studied more easily.

Medical Applications

Complex data are being simplified in medicine, for example, through the use of computer graphics. Since the dawn of computerized tomography (a diagnostic technique that uses X-ray photographs) and the CT scanner, physicians have been using cross-sectional information to assist in diagnosis and surgical planning. Imaging computers can now take this series of cross sections and construct a three-dimensional model of the patient. Surgeons can show patients the probable outcome of reconstructive surgery before they enter the operating room. The computer can interpret photographs and X-rays and provide the plastic or orthopedic surgeon with information on the extent and need for structural correction.

Scientific Applications

Illustrations created by the artist on the computer serve many uses, and the computer itself can create images that clarify and explain. For years, geophysicists have relied upon seismic data to predict where oil might be discovered underground. The sheer volume of data involved in estimating what lies under the earth at varying depths made the job daunting. Graphics computers that can assimilate this data and generate images in three dimensions make this work far more understandable and provide a method for more confident decision making.

ANIMATION

Animation, of course, is nothing new to video and has been used by some producers in nonbroadcast applications. These uses often involve cel or model animation, both of which are complex and expensive propositions.

Cel Animation

Cel animation is perhaps the slowest and most expensive form of animation and has the limitation of a narrow range of options in regard to the appearance of the final product. The name is derived from the acetate sheet, or cel, onto which an object is drawn. Each object in the picture is drawn on one of these clear cels. The background is created on another cel or on another type of art board. Each frame is created by laying all the objects on the individual cels onto the background; the resulting sandwich is photographed.

One principal drawback of this process is the time it takes to produce a sequence of animation. There are typically 18 frames per second of film, and some animation will shoot each cel onto two frames. So, for every second of animation, nine cels must be produced, each showing minute differences in the object as it moves. If more than one cel per frame is being used, for more than one moving object, for example, or if many backgrounds are required, this process gets much more complex and expensive.

Model Animation

An alternative to cel is model animation. The objects in the production are created out of wood, clay, or other building materials. Often the replicas (for example, buildings) are in miniature. Sometimes models of company logos are created for animation. As with cel animation, the process is costly. It is time-consuming to create models. For example, a building is often created with its surrounding city block, and many models are required for the production of the sequence of animation. Sometimes, as in the case of the company logo, the model may not be very complicated, but the process of creating the animation is. It is not uncommon for animation of a model to undergo five or six photographic processes during the production. In video, as we described earlier, we would call these steps "passes." Unlike in video, very little quality is lost as we undergo each "pass," but the time it takes to copy or retouch the photographic product is far greater than when we are operating outside the digital realm.

Computer Animation

The computer can be used as an alternative to these techniques, often with significant cost savings. A computer can replicate scenes and characters rapidly and electronically, thereby eliminating the need to reproduce large portions of moving figures on individual cels. Once copied, only the moving part of the character must be altered.

Even when not utilized to replace the cel technique entirely, computers have been used in conjunction with cels to speed and simplify the process of animation. In some houses, computers are used to create some of the layers for multilayer cel animation, and traditional ink and paint cels are used for the remaining layers. Once again, by speeding the process, it becomes a less expensive proposition.

Tweening and Color Cycling

Many computer systems provide the animator with the capability to metamorphose from one image into another. The process, called in-betweening or *tweening*, reduces the number of cels the animator must produce. If a character spends one-half second raising his arm while the rest of his body remains still, the artist need not produce fifteen cels with progressively changing arm positions. Instead, the computer can be instructed to change the frame of the character with his arm down into the one with his arm up. The artist tells the computer how long the move should take, and the machine does the rest.

In some instances, the desired effect can be created with little motion or moving or pulsing color. Many systems are capable of creating this effect without involving a process as complex as cel animation. Color cycling or paint animation offers a limited range of capability in this realm, but can be accomplished on many, inexpensive systems, which are frequently within the budgetary limitations of corporate video. The process of color cycling and tweening will be described in Chapter 4.

Rotoscoping

Another technique previously relegated to film animation—*rotoscoping*—is also possible on the computer tablet. The process is largely the same: place an image on the screen or a hard copy on the digitizing tablet, trace over it with the electronic pen or brushes, and layer them sequentially on videotape. The recording process is, however, far simpler and less expensive than with film. For example, Figure 2.16 was originally a photograph of a diver. The picture was frame-grabbed into the computer and rotoscoped using a paint package.

Figure 2.16: Rotoscoping. (See Plate 16.)

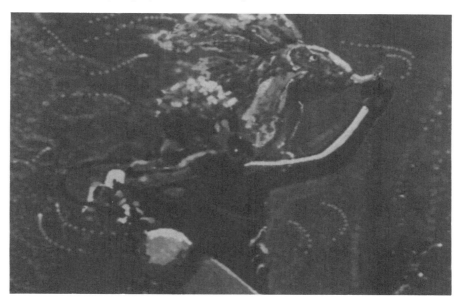

Courtesy Electric Paintbrush, Inc.

Rotoscoping in the traditional sense requires replicating, drawing over, and rephotographing each frame of the film. The computer offers a shortcut in that each frame can be saved in the computer immediately, eliminating the need to rephotograph the frame. Additionally, the artist can often select both the color to use in drawing over the existing image and the "tip" to assign to the stylus. Selections often include the hard-edged line of a marker or ink pen, or rougher lines like crayon or charcoal. Finally, many systems offer the ability to superimpose one image on another. This eliminates the need to make a print from each frame of the sequence to be rotographed.

CHARACTER ANIMATION

Computers are making the movement of drawn characters a much less expensive proposition than cel animation, and, thanks to the progression of software, are concurrently offering the special effects quality of three-dimensional presentation. The use of animated characters offers the producer some opportunities not previously available when outfitted with cameras and recorders alone.

Cartoon Characters

Cartoon characters can be presented as caricatures, with exaggerated speech and conduct. The color can change dramatically in an angry manager's face, and steam can come out of his ears. A stubborn employee can be a mule, and a deceiving vendor's nose can grow like Pinnocchio's. Such hyperbole can be instructive as well as humorous. Since only those attributes the producer wants to emphasize are exag-

Figure 2.17: Phone People. (See Plate 17.)

Courtesy Robert Abel & Associates

gerated, the point being made is more direct, and the lesson can be easier to remember. The speaker need not even be human. (See Figure 2.17.)

Smokey the Bear has been presented on television many times and is more widely identified with fire prevention than any human spokesperson. Some firms have been represented for a long time by a particular print character and wish this company symbol to make presentations in video as well. One of the first examples of computer-simulated human motion for a commercial was generated for the Michelin Man.

Producers in some cases simply wish an abstraction from a human spokesperson. Smokey can speak to viewers as if they were children, and people may not put up their defenses as quickly because Smokey is a talking bear, not a preaching human. When the Canned Foods Information Council wanted to let people know that canning technology would be used to preserve food for centuries to come, they opted against having a person tell the story. Who knows what people will look like in the year 3000? Why worry about it? Instead, they built a robot in the computer, and it did the talking.

Presenting information through a cartoon character is often effective in creating the desired effect. Pediatricians in the Orient (and increasingly in the West) use dolls to assist in diagnosing children's ailments. A child can point out where he/she hurts on the doll without embarrassment or inhibitions. These approaches are not limited to children; the characteristics of animated characters that enable children to overcome psychological barriers can help adults deal with their fears and prejudices as well.

Industrial Uses

Eastman Kodak produced a series of videotapes that portray the most common attitudinal roadblocks to productivity. Each negative attitude was represented by a different tribe of gnomes. *The Gnome Chronicles* was produced to highlight attitudes that defeat teamwork within a company. Michael Perlson, an industrial psychologist who has viewed the series, doubts that people caricatured in the tapes will recognize these traits as their own, but believes the program has value because it involves viewers in a discussion about how their organization works and why some of their fellow workers behave as they do. He believes this discussion should help workers and their organizations to function better.

GRAPHICS AS PART OF THE PRODUCTION

One note of caution is necessary in this overview of the uses of computer graphics. This chapter has covered all kinds of uses for computer graphics systems; some are appropriate for a variety of systems, others require the capabilities of the largest dedicated systems in use. Many corporate producers will find that budget constraints limit them to the use of only a few of these techniques initially. Do not despair. Most of the advances in computer imaging have come from creative answers to problems, and not by bigger dollars for research and equipment purchases. Utilizing a simple graphics computer with other post-production equipment to enhance its capabilities will be discussed in Chapter 4. Such methods can help you get the most bang for your computer graphics buck.

Furthermore, because these applications and effects are possible does not automatically qualify them as desirable for most projects. Computer graphics offer the ability to communicate with visuals not previously available to the corporate producer. They also offer ample opportunity for abuse, misuse and overuse. Appropriate utilization of the technique is half the story behind making them effective. Do not get caught in the trap of using computer graphics to make everything flash and fly though space unless that will enhance the communicative value of the production. One of the reasons Scanimate had such a short run of popularity in television production was that many producers snapped up the opportunity to use it soon after its introduction and used it too frequently. There are producers who have said if they see one more intro with chrome letters flying over a star field, they would lose their breakfast. Many viewers would agree. Content should dictate the production technique utilized, and the corporate producer now has a powerful new tool for communicating that content.

Beyond using computer graphics for their own sake, beware of making a stunning graphics presentation at the expense of the production's budget. One memorable segment within a program can go a long way toward successfully getting a message across, but seldom can an entire program's content be communicated in one segment. One of the fundamental differences between broadcast commercials and corporate communications is the amount of information to be communicated. The advertisement needs to communicate very little: the product, the cost, why you have to have it, and where you can get it. The goal of a corporate piece is far more complex, and many messages must get through to the viewer. Use computer graphics to help communicate those many pieces of information, but bear in mind that each segment is one of many important elements in the entire presentation. Do not sacrifice the communicative value of the entire program for one ''gee whiz'' piece. Keep the graphics budget allocation in proportion to the importance of the graphics segment's message.

3 Two-Dimensional Imaging and the Paint System

We will discuss computer graphics systems roughly in the order of their sophistication and cost. The operational assumption will be that if you are considering incorporating your first computer graphics into a production or considering the purchase of your first system, it is more likely to be a graphic or machine that will represent the least risk (within reason) and find the most general application in your situation. It is unlikely that a video facility's first piece of hardware will be a highly complicated and costly animation system. Rather, a more basic system that will find general use will be the first choice of the video manager. The first application of computer graphics will likely be an introduction or company identification, which can be produced at relatively low cost and will allow the distribution of the cost over many productions.

The cost of producing a computer graphics or animation sequence depends upon the complexity of the image and its duration. A static image, composed of type and a design of solid color, will be dramatically less expensive than a moving aerial view of a computer-generated landscape with realistic shadows and reflections in the lake below. It takes a less complicated system to compose a free-form illustration on the video screen than to build an imaginary environment within the computer and view it through the video screen. Therefore, the less complicated system is the one most likely to be utilized by a producer as a first step into computer graphics. Most of the more sophisticated systems offer many of the same basic functions and build upon them. In the same manner, our progression in discussing different systems and their capabilities will progress from the relatively simple to the very sophisticated.

TYPING AND GRAPHING

One of the most basic advantages of using a computer for any application is its ability to perform functions that are repetitive, time-consuming or uninteresting. In the world of computer imagery, this translates into replacing art cards and generating similar graphics.

Character Generators

As an alternative to live video, producers have always turned to art cards for titles and graphics. With the introduction of character generators, titles and text could be produced without the involvement of the "art department." Recently, character generators have been developed with functions that broaden their usefulness in the video facility. Character generators now are capable of producing results similar to digital effects; text can be typed onto the screen, as well as manipulated and positioned along with logos and background graphics. These systems typically offer a few graphics capabilities and a limited color palette to perform functions beyond those of the simple character generator, but with substantially fewer capabilities than a "full" graphics system. In this chapter, we will refer to character generators in only their most narrow sense (essentially video typewriters, which place letters and numbers on the screen). We will consider more sophisticated machines' graphics systems even though their manufacturers may still refer to them as character generators.

Business Graphics

The next level of sophistication in computer imagery is what has been popularized in the computer software field as business graphics. Designed to render lengthy or complex information in an understandable form, this class of graphics has long been utilized in print and other audio/visual media in the form of bar graphs, pie charts, and so on. (See Figure 3.1.) Many software packages are now available for use with microcomputers and some character generators.

Background Graphics

A more sophisticated use of the simpler two-dimensional systems is the production of backgrounds. Titles, graphs and transitions using digital effects look far more professional when performed over a simple background rather than over black. (See Figure 3.2.) Many video switchers provide solid color backgrounds over which titles or images can be superimposed. More sophisticated designs and patterns are possible on more complicated systems.

Ramps/Gradations

One of the simplest of these techniques is the *ramp*. This is a grading of color from black to solid color, or through shades of a color. The ramp can progress along the entire screen or some portion of it, vertically, horizontally, or at an angle chosen by the operator. Several ramps can be arranged to form a pattern. The ramp is to color over the length of the screen as the dissolve is to video over time.

Figure 3.1: Depth added to simple graphics.

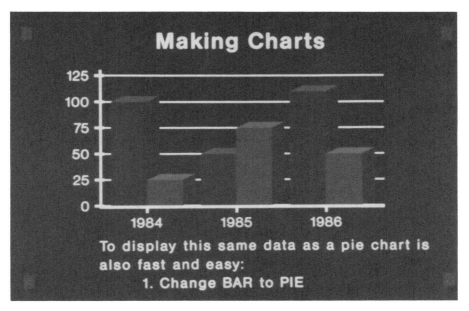

Courtesy Artronics Corporation

Figure 3.2: Type over image.

Courtesy Artronics Corporation

Repetitive Designs

Another popular technique for background graphics is the use of a repetitive design. These are typically done in subdued colors, often with simple patterns. They may be held motionless behind a title or window of video, but can also be made to move. Motion backgrounds are very popular in broadcast, especially when used in conjunction with moving titles. Utilizing some of the airbrushing and shading features of paint systems, the background pattern can also be made to appear three-dimensional. One example is the generation of a simulated ''diamond plate,'' inspired by the material used on truck running boards.

These are, of course, only a few examples of what can be done with even the simplest two-dimensional imaging systems. We now turn to the technology of two-dimensional systems.

PAINT SYSTEM ARCHITECTURE

Graphics systems designed to create free-form, two-dimensional images rather than defined characters are called paint systems. They treat the video screen as a canvas onto which images are drawn or composed with an electronic pen or brush. They are fundamentally different from the more sophisticated modeling systems, to be discussed in Chapter 5, in that they create images not by generating a mathematical construct of an object, which is subsequently viewed through the video screen, but by manipulating the raster directly.

The Frame Buffer

The invention that made paint systems possible was the frame buffer, still an integral component of all computer graphics systems. As described in Chapter 1, the frame buffer is a dedicated amount of computer memory capable of holding one frame of video. The buffer is used for temporary storage of an image. Alternatively, it may be a workspace, holding an image as it is being created or manipulated. It may be a dedicated area of memory within a computer, or a separate device.

The frame buffer's original application involved accepting pixel information from a graphics processor and updating the video screen thirty times per second. In early applications, it was little more than a storage device for images between the processors and the CRT. Turning the frame buffer into an active part of the system opened the opportunity for the development of today's illustration systems.

The man responsible for this was Richard Shoup. Working at Xerox's Palo Alto Research Center, he viewed the frame buffer as a device that could be addressed directly. We have defined a pixel as a memory location in the computer, which contains information about the color and brightness of a specific point of light on the video screen. In systems prior to Shoup's, the operator used the computer to construct an element of a picture, a line or geometric shape, for example, and the computer assigned the color of that design element to the appropriate pixels for output

Figure 3.3: Digitizing Tablet

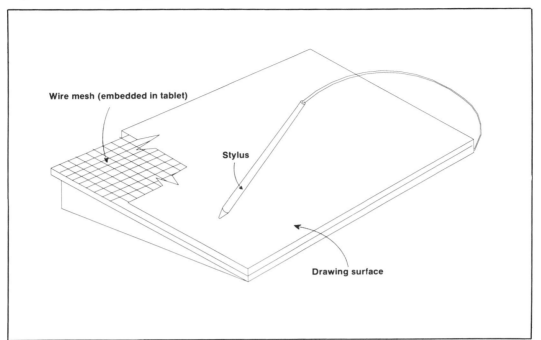

to the screen. In Shoup's system, the operator assigned values to pixels directly, and the frame buffer output them to the screen. There would no longer be a need for complex programming to generate a simple image; the technician could simply color pixels to form images. The product of this initial research was Shoup's Superpaint system, which was used to create television's first raster-based graphics in 1976 for the PBS series *Over Easy*. Development of the modern paint system is largely credited to Alvy Ray Smith, who developed his program, *Paint*, while working on Shoup's team (which had by then moved to the New York Institute of Technology or NYIT) in 1975.

The basic hardware defining today's paint systems consists of the computer's central processor, the frame buffer, which stores the image and in which the image is altered, storage devices for maintaining pictures after processing, and input devices, which allow the artist to create and manipulate images. It is these input devices that revolutionized the graphics system by removing the principal barrier between the artist and the computer. The fundamental input devices that are components of today's paint systems enable someone with virtually no familiarity with computers to begin creating illustrations quickly and easily.

The Bit Pad or Digitizing Tablet

The fundamental tool for computer illustration today is the bit pad or digitizing tablet. (See Figure 3.3). The most common form of the digitizing tablet looks like a blank pad or writing surface. Embedded in this pad is a mesh of very fine wires.

Figure 3.4: The artist can select from a number of brushes on most paint systems.

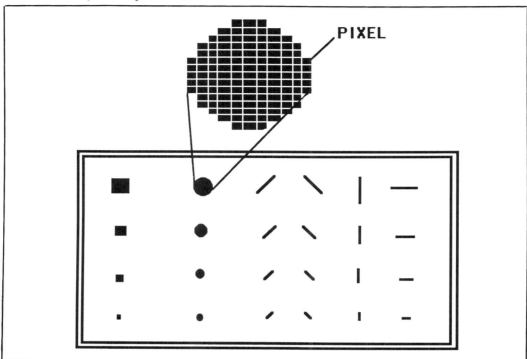

The intersections of these wires correspond to the locations of each pixel on the screen. Attached to the pad most often is a stylus, resembling an electronic pen, connected to the pad by a wire. Touching the stylus to the pad induces a weak electric current in the mesh, which tells the computer the location of the pen on the tablet. After selecting a function, the artist can indicate changes to specific pixels by passing the pen across the pad. Today's more sophisticated tablets are surprisingly accurate, some providing accuracy to a thousandth of an inch. They can transmit information about activity on the pad up to 200 times per second, permitting interactive image manipulation.

Selecting Brush Shapes

To use this tablet, the artist first chooses a "brush" for the stylus. This is usually a matrix of pixels, which defines a shape such as a circle, stored in a shape table for selection by the operator. (See Figure 3.4.) One of the available brush styles is chosen by touching it on a display with the stylus. From then on, whenever the stylus is touched to the pad, a copy of that brush shape is copied into the frame buffer.

Selecting Colors

The artist selects a color the same way. A subset of the available colors is presented somewhere on the screen as a row or grid of color boxes, the "palette."

The artist selects a color by touching the appropriate spot on the grid with the stylus, almost as if the brush were being dipped into a jar of paint. Once the stylus has been defined by a certain shape and color, the artist may paint freely on the tablet. The positions of the stylus as it glides over the surface of the tablet are relayed to the computer, and a line appears along this path on the screen.

Although the size of the palette (the number of colors that may be selected at any one time) varies from one system to another, most offer over 16 million colors. On many, though, only a portion of these colors may be displayed simultaneously. The size of the palette can range from 8 to 16.7 million colors. Most systems offer either 256 or 16.7 million. The attribute that dictates how many colors may be described on the screen is depth of color, and it is expressed in bits per pixel. The more bits devoted to describing each point of color on the screen, the more colors there are that may be displayed at once.

PAINT SYSTEM STRUCTURES: LEVELS OF COMPLEXITY

The basic structure of paint systems is fairly uniform. While the method of manipulating an image is relatively standard, a variety of hardware is available to accomplish the drawing and painting tasks. The capabilities of an individual system tend to parallel its complexity and cost, but, depending on the requirements of a job and the constraints of a budget, a surprising range of options is open.

Personal Computer Graphics Systems

A significant advance in hardware, which allows the corporate producer access to inexpensive graphics, is the increasing sophistication and flexibility of the personal computer. Plug-in peripheral circuit boards enable PCs to produce near-broadcast-quality images with less than $10,000 worth of equipment. While the components of a computer system are called hardware and the programming that makes it run is software, these plug-in additions are called *firmware*. Firmware provides specialized circuitry to accomplish a specific function and comes with some programming indelibly inscribed into its memory.

The decrease in the cost of computing power, the increase in the resolution and color capabilities of PC software and the broadening familiarity with PC architecture among video engineers are some of the reasons for this increase in availability. The power of some microcomputers, in fact, is so impressive that many of the broadcast systems designed as dedicated graphics and animation computers are based in micros.

Most of the firmware available as plug-in additions to the micro bring the PC to the level of a simple graphics system, but some can be quite sophisticated. Number Nine Computing in Cambridge, MA, offers a video board for the IBM Personal Computer that features pan, zoom and RGB (red, green and blue) or NTSC (National Television Standards Committee) input and output. In addition, nearly all board manufacturers offer a video ''frame grabber.'' Instead of creating an image from

Figure 3.5: Mouse.

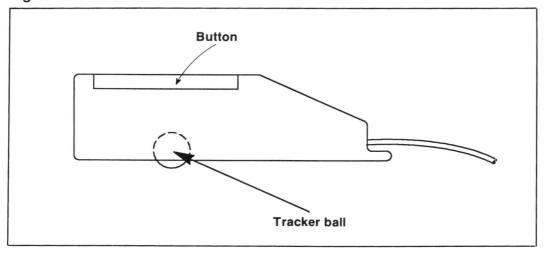

scratch, this option enables the computer to capture a frame of video from some other source like a camera, manipulate it with the paint or graphics program and store it.

Operation of a PC Graphics System

The operation of a PC graphics system differs a bit from its big brother, the dedicated graphics workstation. The artist sits in front of a single screen, which is often divided between image and menus. The image under construction is presented on most of the screen, and command prompts are positioned along the side or bottom of the screen in abbreviated words or icons. An icon is a little picture that indicates an available function. It describes a menu item in graphic form rather than in written form. For example, a menu of words may include a Delete File. The icon equivalent (as anyone who has worked on an Apple Macintosh has seen) would be a garbage can.

Image composition can be performed from the keyboard or with a peripheral device, such as a mouse or a puck. Unlike more sophisticated systems, a digitizing tablet is often not part of a simple PC system. A mouse is a box containing a rolling ball or tracker, which is exposed to the table surface underneath. Moving the mouse over the surface of the table moves the ball inside and relays the appropriate information to the computer. (See Figure 3.5.) Drawing a line, for example, can be accomplished by selecting the appropriate drawing mode in the program, positioning the cursor over the spot where the line is to begin, and moving the mouse to create the line. Move the mouse to the right, and the line grows to the right; move the mouse to the left, and the line grows to the left, and so on.

Personal computers most often store images on floppy disks or hard disks. Access time varies with the type of storage medium and operating system, but most

Figure 3.6: PC-based system.

of these systems are capable of recalling only one image for manipulation or transmission at a time. Real-time animation on such systems is not possible (yet).

Working with these computers is almost identical to working with your PC in any of its other, more general functions. The system is little modified from its normal role. One peripheral board (the firmware described above) and perhaps a drawing aid are all that differentiates it from the machine on the secretary's or bookkeeper's desk.

The PC-Controlled System

As we progress up in levels of sophistication, next is the PC-controlled system. While the previous systems were microcomputers outfitted with firmware to produce graphics, PC-controlled systems are dedicated systems that utilize PCs as their brains and sophisticated software to control their functions. The micro assists in image construction and acts as controller for the other peripheral units, which process, store and transmit the images. Many broadcast systems are constructed this way. The system configuration may present the graphic artist with two screens: an image screen and a data screen. (See Figure 3.6.)

The image screen contains the picture being manipulated and some information about the mode currently in use. Often this takes the form of a color palette and a menu. The menu shows the artist which functions are available and offers an easy way to select one. Position the cursor over a function and press the enter key or the button on the mouse or pen and the function is activated. Choosing many of the more general functions yields a new menu in place of the last to further define the desired function. The menu typically consists of a series of abbreviated words or icons.

The data screen displays information about the system and may give prompts for the function being performed. This screen is used by the artist to determine such information as remaining computer memory, disk contents, image names and system status.

These systems have additional storage capacity, necessitated by the higher complexity of the images produced and the increased requirements for access speed. Rather than relying on a mouse, they are commonly outfitted with sophisticated digitizing pads for image entry.

Supermicro- or Minicomputer-Based Graphics Systems

Finally, at the high end of the two-dimensional packages are the systems based in supermicro- or minicomputers. These systems are frequently assembled integrally with a desk to become a complete workstation. They may be equipped with large-capacity storage media (such as 10 megabyte floppy disks), offer internal signal processing equipment and VTR control, and may even support more than one artist at a time. These systems are usually found in post-production facilities, but, no doubt, some corporate video departments have them also. They are similar in function to microcomputer-based systems, but offer more and faster storage and many special features.

COMPUTER PAINT SYSTEMS IN POST-PRODUCTION

Computer paint systems have applications in both the offline and online settings of the post-production process. They are operated independently when creating two-dimensional illustrations rations for a production. Titles, graphs, key frames and other illustrations are created and stored on whatever magnetic media are available to the computer for recall and incorporation into the production later on.

More advanced machines can generate some effects, like color cycling or limited animation, internally. In such instances, the system may be connected to a VTR. Those effects that the computer can perform internally are set in motion and captured onto videotape in real time, later to be edited into the final production.

Finally, the graphics system is used as a source of video during editing. It fits into the system in this case in the same way the slide chain, character generator, or DVE (digital video effects unit) do. The editor may, for example, cut away from a tape of a speaker referring to a growth in profitability over the last few years to a graph of the increasing annual profits, and back to the speaker.

PAINT SYSTEM FEATURES

Sophisticated paint systems offer many more features and capabilities per option than simple paint systems. Note, however, that a purchasing decision based solely on how many features a system has may not make sense. A more thorough discussion of contract or buy decisions appears in Chapter 7. The attributes mentioned in this chapter offer some points on which to compare systems, orient you to some of the more common terminology of paint systems and indicate the relative worth of various features.

Number of Colors

The simplest systems offer up to 256 colors for titling and simple illustration. Although potentially useful for replacing art cards and other uncomplicated graphics tasks, a color selection this limited can do little in the direction of rendering video images. At the other end of a wide spectrum, most broadcast paint systems offer up to 16.7 million colors, although only a small portion (256 or 600, for example) may be available at any one time on the screen. The power of this variety of colors is realized in rendering the shading for realistic images. For example, images that include perspective or subtle shadows can only be created by utilizing the infinitesimal differences between shades of the same color. A limited palette would be either incapable of creating shading or would leave telltale bars of color along a surface, rather than a smooth gradation of color. Again, for most graphic or technical illustration, this range of color may not be necessary.

Depth of Color

By depth of color we do not mean saturation or brightness. In graphics paint systems, depth refers to the amount of information devoted to describing each point of color on the video screen. Computers store information in a series of units called bits, which may take on the value zero or one. For the CRTs that originally presented vector graphics, only a single bit was necessary to communicate the value of a point on the screen, since the point could only be black or white. For raster-scan images, more bits are required to describe individual pixels because various shades of grey are desired. In color systems, it is necessary to describe hue as well as intensity in order to construct more intricate images. An increase in bits dedicated to describing a single pixel will result in higher accuracy in shading and definition. In 8-bit systems, we say each pixel is 8 bits deep. An 8-bit system designed for television graphics might, therefore, render an image that is 512 x 512 x 8 bits, reflecting 512 lines on the screen by 512 pixels per line by 8 bits of information per pixel.*

---

*This concept is not to be confused with the system speed, characteristically expressed in bits. The IBM PC is a 16-bit system. That means its components communicate in words that are 16-bits long; it exchanges information between its operating units 16 bits at a time. The Apple II, on the other hand, is an 8-bit system. IBM is not necessarily a better machine; it is just twice as fast when it is performing operations where the different parts of the machine are communicating with each other.

Draw Functions

Every computer graphics package can draw straight lines. The big difference between machines is how the task gets done. Some systems require the artist to plot the endpoints of the line, and then the system creates a line between them. A more common method is the "rubber band" function. In this process, the artist designates the starting point of the line by selecting the appropriate function, touching an electronic pen to the digitizing tablet, or pushing the button on a mouse. The line is then "pulled out" of this point as the artist moves the pen or mouse. This way, the finished line is visible before the artist dedicates it to a fixed position. Geometric shapes can also be constructed with the rubber band technique, and most systems take advantage of this. Squares get "pulled" out of a single point by defining diagonal corner points; circles are formed by defining a square into which the circle is inscribed.

Curved lines are a bit more complex to render, and when comparing systems, it is worth noting what method is most comfortable for the artist who will be using it. There are methods that use a limited number of control points to define curves which will be smoothed between them. More sophisticated techniques of rendering more realistic or smoother curves will come at a higher cost, and may be more difficult to master.

Grids

Grids and movable axes are often available to assist the artist in aligning picture elements on the screen. A grid can be called onto the screen, objects positioned, and the grid removed without disturbing the image on the screen. Of course, the grid has become a popular picture element on the screen, so the artist may elect not to remove it after all. Removable grids are used to align type on the screen, to center or accurately position elements of an image on the screen, or to create even spacing between elements.

Positioning

Another useful function is flexible positioning of picture elements. Many systems allow the operator to outline or enclose an image and adjust its placement by moving it to a selected area on the screen. If a design turns out bigger than anticipated, when the client decides he wants his logo in the upper left corner rather than the middle of the screen, or when the picture drawn is not properly centered, the positioning function allows the artist to revise the size and placement of the image with little difficulty.

Replication

In the process of replication, an element on the screen is selected, as in the position function, and duplicated elsewhere on the screen. For a repetitive image background, this function allows the artist to draw the pattern once and copy it many times on the screen. When a pattern is created by drawing one small element and replicating it many times on the screen, the technique is called *rubber stamping*.

Some systems allow an individual element of an image to be saved separately from the entire picture. The artist defines the element as above, but duplicates it into memory instead of onto another part of the screen. In this way the artist can build a scrapbook of frequently used images. Company logos, trademarks, electronic symbols and icons can all be saved and called into a new image at will.

Scaling

Size changes of existing images are made possible with scaling. For example, if an artist constructs a company logo but decides that it is out of proportion with the other information on the screen, scaling will enable the artist to reduce or enlarge the image until the desired proportional size is achieved. By combining this function with replication, a complete image can be constructed from a single element in a very brief time. For example, the artist is told to graphically express the growth in company earnings for the past five years using a single, static image. He decides to pull a company logo from the image file and places the relatively small element on the left side of the screen to represent the first year's earnings. During the second year, revenues rose 20%, so the first logo is defined, replicated and scaled up to be 20% larger than the first and placed to the right of the first logo. The next year, earnings rose another 12%, so the artist enlarges the second logo by that additional percentage and then repeats the process for each successive year included in the graph. At the end of the five years depicted, he types the appropriate years under each logo and has a finished graphic.

Mirroring

Mirroring enables the operator to create the left-to-right or top-to-bottom reverse of an image. The computer's idea of a Rorschach test, this technique allows the simple creation of symmetrical pictures. One possible application of mirroring is the creation of a shadow or reflection of a company logo in a simulated three-dimensional illustration.

Rotation

Rotation enables the artist to change the direction of an object, in effect spinning the object on an axis. This can be useful when, for example, the operator has created a company logo that sits on a horizontal line on the screen, and the client wishes the logo at a 45° angle in the corner of the screen.

Radial spreads are available on some systems to create images out of lines or points by rotating them around a point or axis, leaving the trace or a tail in its wake.

Figure 3.7: Paint Functions. Created on Artisan.

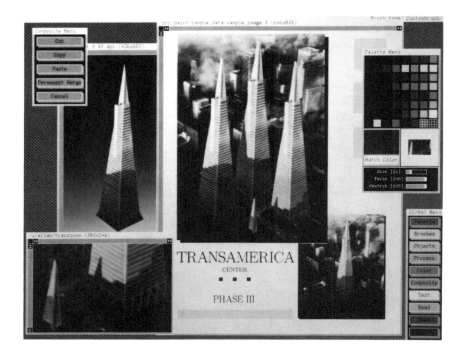

Courtesy Media Logic Incorporated

Paint Functions

The principal result of the draw functions is to create an image composed of lines, similar to penstrokes. The paint functions provide the artist with methods of coloring those images and of creating more free-form pictures.

One feature of almost any graphics system is the ability to select from a variety of brushes. Instead of manipulating previously defined image elements, painting, like drawing, allows the free-form creation of images. Any number of curves or shapes is possible above and beyond the better-defined geometric shapes created by the draw functions. In the most basic systems, the artist selects the desired width of a line. As the machine becomes more complex, the edge of the line can also be adjusted, producing a range of definition from a hard line with crisp edges to a very soft line that has solid color in the middle and quickly fades off into the background color as the edge is approached.

In Figure 3.7 the TransAmerica Building is used to demonstrate a sophisticated application of some basic paint functions. The photograph of the building was frame-grabbed into the paint system, as shown in the bottom right corner. The building itself was cut out, as seen in the window on the left. That image was replicated back into the original photo twice, resulting in a picture of three TransAmerica buildings, the illustration in the center. Just the top of one building was cut out as well and is shown placed back into the center picture as the small spire in the foreground. The lower right corner contains an image that has been zoomed in for closer inspection. The illustration also shows some examples of menus and a palette.

Airbrushing

From this soft brushstroke, we progress to another common technique, the airbrush. Airbrushing can produce lines or strokes as well-defined as a soft brushstroke and progress to very wide patterns, much like using a real airbrush on paper. Areas can be airbrushed heavily to create solid color, or may be brushed quickly to lightly tint or highlight an object. In the most sophisticated systems, opacity of the airbrushed line can be controlled by the tilt of or pressure on the electronic pen. The artist may be able to use masks, stencils or friskets to create sharp edges on the strokes.

Anti-Aliasing

Since the video screen is comprised of a matrix of square pixels, true, smooth diagonal lines are impossible. Furthermore, the shallower the angle of the desired lines, the more pronounced the problem. The way a computer adjusts for this shortcoming is by creating an ''alias'' of the diagonal line, lining up a series of discreet line segments approximating the path the diagonal line would follow. Looking closely at the diagonal, the viewer can detect this staircase of line segments, often referred to in video parlance as ''jaggies.'' (See Figure 3.8.) Many computer graphics systems have software to help correct this problem, a feature called anti-aliasing. The computer finds each step in the line and inserts a pixel of a color midway between the line and the background. (See Figure 3.9.) This softening of the line edge actually results in a sharper-looking diagonal, thanks to the compensatory mechanisms in the visual center of the human brain. Obviously, the higher the resolution of the system, the more effective the technique.

Frame Grab

As mentioned earlier, most computer graphics systems have the ability to ''grab'' (capture and store) a frame of video taken from another source, such as a camera. The differences among systems lie in their varying capacity to interface with other pieces of video equipment and the amount they can grab within a given time. The most common ability is the capture of a single frame of video from a camera or VTR. There are systems, however, that can record a series of frames and store them separately to be manipulated individually. The more sophisticated graphics machines interface with digital frame store devices, enabling the artist to select images out of the frame store for adjustment. When approaching the problem of retouching video frame by frame, this combination of equipment in invaluable.

Figure 3.8: Aliasing.

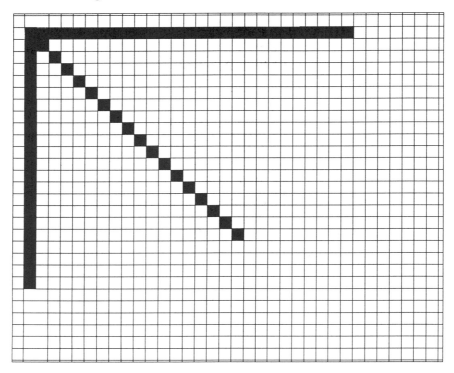

Figure 3.9: Anti-aliasing.

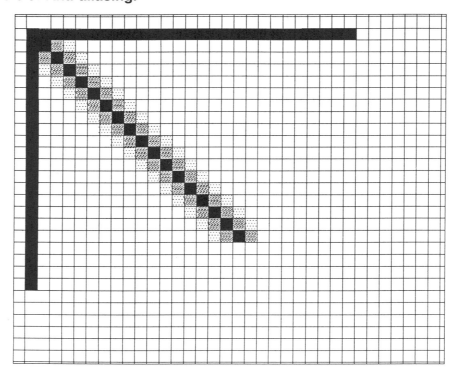

HARDWARE FOR PAINT SYSTEMS

Input

Several techniques for inputting data to manipulate images have been discussed in this chapter. It bears mention here, though, that the method used to communicate with the computer is a very important consideration when deciding on a purchase. The digitizing tablet, although not always available on the less-expensive systems, helped to eliminate the barrier between computers and artists by making computer graphics composition similar to using pen and ink. If your needs dictate acquiring a system without a tablet, make sure the artist who will operate it is comfortable with using a mouse or puck.

Storage Media

The smallest systems utilize floppy disks of the $3\frac{1}{2}$-inch or $5\frac{1}{4}$-inch variety. These disks can store limited numbers of illustrations and may take some time to read into memory, depending on the operating system of the computer. More commonly, a hard disk will be required to run all the software required and to provide working storage for the images. These units can be as small as 20 megabytes and can be installed in the central unit of a personal computer. More sophisticated systems use larger floppy disks, capable of storing 10 megabytes or more. The Bernoulli Box by Iomega is one example. These systems often come equipped with a peripheral hard disk (meaning the storage disk is in a case separate from the computer). The high-capacity floppy and hard disk are typically all that is required for two-dimensional paint systems. If the purchase of a system is being considered, however, do not think in terms of how many peripherals you intend to purchase for storage. Think in terms of how much the machine can perform without having to access a storage device, or of how often the operator will be called upon to change disks. Base your decision on how often the operator will be distracted or delayed from creating imagery because new disks must be inserted.

The most sophisticated systems actually split the job of image creation into two computers. The operator sits at a workstation and describes the image on a ''front-end'' computer. This computer is equipped to provide interactive support for the input devices and allows the artist to compose an image quickly and efficiently. The front-end, however, lacks the computational power required to perform the complex functions involved in creating shiny surfaces or light sources. These responsibilities, then, are turned over to a ''rendering engine,'' which calculates the specifics of each frame of video and stores or outputs the finished images. This system arrangement is almost never used to create two-dimensional imagery, but will arise when we discuss three-dimensional and animation systems in Chapter 5.

Output

On the smallest personal computer systems, the only way to get an image from the monitor to tape is to tie into the video output of the computer. This poses problems in that the signal designed for the computer's screen may be nonstandard video, of low resolution and lacking information your equipment requires. With the proliferation of PC firmware, however, this is no longer a concern. Most plug-in boards offer RGB or NTSC composite signal outputs, and the larger dedicated machines have further signal processing capabilities to facilitate interface with other video equipment. If it is important for your graphics system to be synchronized with the rest of your equipment, some systems have the option of genlock. Machines are frequently available with video equipment control abilities to allow the artist to record images onto tape without leaving the workstation. These capabilities will also be examined more fully in the discussion of animation in Chapter 5, where machine control becomes a more important issue.

Alternate Forms of Output

How important are alternate forms of output? Some of the more sophisticated modeling systems are available with film recorder or optical printer output, but even some paint systems are available with a means of capturing images from the system onto slides. In fact, there are some fairly complex systems designed for slide output used for presentation graphics. Color printers are available for some systems as well. Although it was an early goal of Ampex to do away with "dog-eared files of illustrations," some artists prefer to make changes by scribbling on a hard copy. If the computer system is to be used for storyboarding, this output is, of course, a necessity. The existence of a paper copy of the graphic also allows an easier route for obtaining approval from a client before incorporation into a production.

CONCLUSION

You should now have some idea of what a paint system can offer and some of the important features to look for when evaluating a system for purchase. Still images created on a paint system can add substantially to your audience's interest in a program. Many facilities keep their paint systems busy all the time creating such images. For the creative producer, however, more spectacular effects can be created for a production using these same systems. Some three-dimensional motion and simple animation can be introduced without having to hire or acquire a more complex machine. It is to these possibilities that we now turn.

4 Simple Animation: The Paint System Meets DVE

Having covered the basics of two-dimensional paint systems, we can take the images created with those techniques and make them more interesting by adding the dimension of motion. The most obvious approach is to treat your illustration the same way the video recorder treats a still image. The recorder views a frame as a single piece in a long series, which, when displayed in series, becomes a program. We can create a similar series of individual drawings, which when presented one after another, becomes an animation sequence. There are, however, far more simple ways of inserting some motion graphics into a production, which are not as time-consuming or complex to produce. These techniques utilize the special effects generators and other equipment in the post-production suite. In this chapter, we will describe a number of these techniques. We will give some suggestions on how to combine the paint system with other equipment, which you may already have at your disposal, to add some motion to your two-dimensional images.

ANIMATION DEFINED

Animation is a series of pictures viewed in rapid sequence to give the illusion of motion. Film animators create this effect by drawing pictures of characters on a series of cels. A cel (named after its original material, cellulose) is a clear plastic sheet onto which is drawn the subject of an animation segment. Once the subject is drawn and colored, the cel is placed on a background. A cel is created for each frame, or two, of animation. On each, the subject changes incrementally. The rapid projection of each of the photographed cels makes it appear as though the subject is moving. This is a simplified definition of cel animation, and many products of animation are many cels thick, and not all are ink-and-paint images. The process, however, is roughly the same.

55

Using the Computer Graphics System to Create Animation

You could do the same with a computer graphics system, making and recording drawings sequentially on film or tape. In fact, a fair amount of animation is generated in this fashion or by using a hybrid of this technique. If you purchased a graphics computer to save time, however, it would probably be counterproductive without some assistance from additional technology. If your service or art department does not currently produce frame-by-frame animation, it would appear that purchasing a computer system to begin offering this service would make productions uniformly more time-consuming to produce. The computer can, however, be used to create some eye-catching and illuminating animation without resorting to a digital version of cel technology. Let us take a look at what options there are, within the computer and by joining it with other equipment, to create simple two-dimensional motion.

COLOR CYCLING

The simplest method of producing animation is available in most of the 8-bit paint systems because of their utilization of look-up tables. A look-up table is, simply put, a reference book for the illustration on the screen. Each field of color is stored in memory with a reference number. This number refers to a position on a palette into which the artist has "poured" a color.

When the computer brings a picture out of memory, it comes out in a paint-by-numbers version. The computer then acts as artist by taking colors out of numbered pots, which are positions on the palette, and filling in the numbered fields with color. Color cycling involves only the manipulation of the location of colors within the palette. Remember that the 8-bit paint system is a color map system—it is palette dependent. The computer takes the numbered colors on the palette and places them into similarly numbered fields in the image, as an atlas maker would fill countries on a world map with various pastel colors. The colors that define the image depend on the current position of colors in the palette matrix. Except areas on the screen defined as the background color, pixels, which comprise an image, have a one-to-one relationship with the palette position from which they are defined. Whatever color appears in or is assigned to that palette position will be taken by pixels defined by that position.

Since the appearance of the illustration depends upon what palette colors are assigned to various positions on the screen, filling those "pots" on the palette with different colors will make the image on the screen appear different. The illusion of motion can be created by changing the color assignment of positions on the palette over a period of time. This technique, called paint animation or color cycling, can be broken into several discrete steps. First, draw a series of successive, non-overlapping images, each defined by successive palette positions. Change all but one to the background color by assigning the background color to those palette positions. Then successively step that group of colors, which constitutes the object, through successive palette positions. Thus, you will be effectively "turning on" each image in succession. The object will appear to move. Color cycling allows the artist to take any color or group of colors and instruct them to successively occupy each position on the palette. In its simplest form, consider the palette as a wheel, and its color values positioned on its perimeter. Spinning the wheel past the color values allows each color to be associated with each position on the palette, one position at a time. The speed at which this wheel spins can be controlled by the artist.

Cycling a single color provides a simple example that demonstrates the process. Take the example of a bouncing ball. Define a ball as a circle of pixels in the upper left corner of the screen, which can be any color designated by the first position on the palette. Just below and to the right of this, draw another circle of pixels and define them as the second position on the palette. Another increment down the screen and again to the right, define a group of pixels as the third palette position. Continue this process to the bottom center of the screen and back up again to the upper right corner. You have now defined the path a bouncing ball will take and assigned each still image with a successive palette position. If you color the first position on the palette red and change all other positions to the background color, you will see only the ball in the upper left corner. Move the red to the second palette position and replace the first with the background color, and only the second ball position will be visible. By rapidly cycling the red through all positions on the palette, it will appear as if a ball is thrown in from the upper left corner and bounces into the upper right. (See Figure 4.1.) If these positions had been defined vertically on the screen instead of from one side to the other, continually cycling through the palette would make the ball appear to bounce up and down.

If your palette goes from white through successively darkening shades of pink and to red and you used it to create a ramp from the top of the screen to the bottom, cycling through this palette would result in a red wave traveling vertically up the screen. It would continue to make that trip as many times as you cycled through the palette.

Another example would be to construct an arrow colored by successive grades of red from white to pink to red and back several times. Cycling through this same palette would result in a pulsating arrow, where the color appeared to move in waves along the length of the arrow.

This technique has been used to simulate such things as flow in industrial illustration, or the transmission of radio waves from a drawing of a satellite to an earth station.

A more complex way of using this technique is to define one or more picture elements as groups of palette positions and cycling these groups through part of the palette. This cycling range would be filled with the background color, except for the current screen position of the moving objects. Parts of objects can be made to move this way. Take, for example, a person waving. Reserve a portion of the palette to define the arm that will move and paint the rest of the person as a stationary image. Define the different positions the waving arm will assume as successive positions

Figure 4.1: Color Cycling.

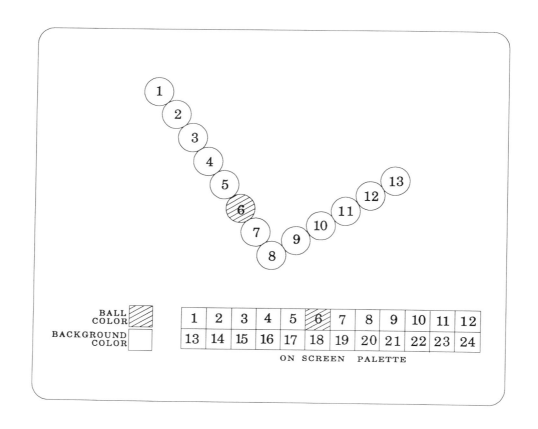

on the palette. Cycling through just the part of the palette that defines the various arm positions will make the person wave. This is, perhaps, not the easiest process to describe, but it is one method of utilizing a very simple paint system to generate some eye-catching animation.

Look-Up Table Animation

A variation of this technique is look-up table animation, in which the image is altered again by different colors successively occupying positions on the palette. In this case, however, the progression of colors is not sequential through the palette. New colors occupy each position on the palette according to a series of tables input to the computer by the artist.

With the dawn of very large-scale integrated circuit (VLSI) technology came paint systems that offer artists the computer's entire color capability on the screen at once. VLSI technology makes possible a frame buffer that is 24 bits deep; that is, each pixel is defined by 24 bits, 8 for each of its constituent colors, red, green and blue. Since 24-bit systems do not have to make use of color look-up tables, animation through color cycling is not possible. There are paint systems available (for example, the Aurora/220 Digital Videographics System), though, that can be switched between 8- and 24-bit modes so that this capability is not lost.

COMPUTER-CONTROLLED DEDICATED PAINT SYSTEMS

Plane Animation

Some systems are now available with software that simulates camera moves or digital effects, such as pan. One such option offered by some manufacturers is multi-plane animation. The computer displays two or more images simultaneously, a foreground image and a background design. The artist may then command the computer to move one image across the other. By recording this move as it happens or sequentially recording a long string of very small moves, you can create the illusion of digital effects without needing them to produce the sequence. This technique, previously found only in more expensive three-dimensional systems, is coming to bit-mapped systems. Although computationally inefficient, newer systems are capable of processing these routines without committing producers to using more expensive equipment.

One of the more useful developments in computer graphics is software that can interpolate between images. What makes cel animation so expensive is the fact that each frame must be drawn by hand. One second of motion can take up to 30 separate drawings. Computers simplified this process tremendously by allowing the artist to copy an image quickly and make changes only to those elements of the picture that were moving. The process, however, remained quite tedious.

Tweening

Now, there is software that performs interpolation between images, a process called *in-betweening* or *tweening*. Give the computer images of a ball in the air and a ball hitting the ground. Tell the computer how much time it has to complete that move, and it gives you back a series of images depicting the ball falling. Have the computer interpolate the images in the reverse order, and it shows a ball climbing into the air. Repeat the series several times and you have a bouncing ball to follow. This technique evolved out of film animation's in-betweening process. Animation specialists would create *keyframes* of action on cels. These keyframes would correspond to critical events in the execution of a move. If the scene depicted a character throwing a ball, for example, the keyframes would illustrate the character with an arm drawn up near the shoulder, ball firmly in hand, the arm partially extended and slightly forward of the shoulder, and the arm fully extended in front of the character releasing the ball. Teams of *in-betweeners* would draw and color enough intervening illustrations to make the sequence, when photographed, appear as a smooth motion.

The basis for today's computerized in-betweening started at the New York Institute of Technology when Alexander Schure, the chancellor of the school, saw the potential for computer graphics to accelerate this process dramatically. He went to Utah and recruited Edwin Catmull and Garland Stern and purchased some of the equipment being developed there (an 8-bit frame buffer being developed at Evans & Sutherland, a company doing government contract work). At NYIT Catmull designed a program called *TWEEN*, which was an attempt to automate much of the rendering between keyframes. Because the algorithms to replicate many of the tricks animators used in their craft had not yet been developed, and partly due to the animator's tremendous distaste for the system, Catmull's program did not meet rapid or wide commercial acceptance. If did, however, form the basis for today's tweening capabilities.

Tweening can be exemplified by using the bouncing ball example referred to earlier. The artist would draw keyframes of the ball in the upper left corner, the bottom center and the upper right corner of the screen. The amount of time the ball should take to move from one position to another would also be fed to the processor. The computer would then fill in an appropriate number of intervening frames to simulate the motion and record the sequence onto a video recorder. This technique can be used for far more intricate images than a moving ball. The preceding, however, is a good example of how a lot of the time-consuming and boring aspects of animation can be filled in by the computer.

The cartoon industry, ironically, still has not widely embraced the idea, due in part to the difficulties encountered in Catmull's research.

Quite a number of animation tricks did not translate well into computer animation when *TWEEN* was written. Motion blur, the loss of detail an animator introduces to convey an object's high speed, for example, was not part of the program. Another problem was that the computer simply showed things too perfectly and the product ended up without the character of a traditionally produced cartoon. When a baseball player hits a ball, the ball becomes ellipsoidal at the moment of impact and then becomes round again as it flies off the bat. An animator gets around this by never showing the impact, but much time was spent in the development of TWEEN in attempting to realistically portray this momentary deformity.

Another obstacle was the animators' resistance to computerization. The group of animators hired by NYIT to help Catmull in his research hated the system. The new process was drastically different from the ink-and-paint techniques they were used to and held the possibility that it would replace the traditional animator.

Innovative Animation Techniques

Hanna-Barbera, with a long history of investigating the possibilities of computer assistance in cartoon production, has developed other techniques for creating cel-style animation digitally. One process they use is a fascinating control algorithm for a paint system. Described briefly, characters' parts, both outline shape and color, are defined in databases. The databases contain almost all possible orientations of each character's limbs, expressions, and so on. When producing the animated segment, parts are assembled hierarchically by the computer according to previously defined rules (i.e., eyes and nose are roughly in the center of the face, the head is always above the torso, and so on). The computer changes the orientation of the parts, or replaces parts, according to the motion the character is performing. By compositing a character's moving parts over a background, perhaps even with another character's pieces being assembled the same way in the same frame, the computer generates a frame much in the same fashion the animator layers cels over a background. The big difference is the speed with which this process can be accomplished by eliminating the need to constantly redraw each cel.

Techniques like this have enabled computer programmers to offer animation software packages to the users of many paint systems. Just as there are firmware additions for personal computers, some simple animation programs are also available for these inexpensive machines. Many of the simpler packages rely on techniques like in-betweening to produce two-dimensional animation.

Motion Cycling

One of the most efficient uses of tweening is moving static images around a screen. That is, moving an object without changing it. Two frames are defined, one with the object to be moved in the starting position and one with the object in the ending position. The program fills in all the intervening frames.

The repetition of this move, or a series of these moves, is called motion cycling. Going back to our example of the bouncing ball, think of the ball bouncing straight up and down instead of from corner to corner. The computer can in-between the ball falling and rising after being given the key frames of the ball in the uppermost position and the lowest position. To have the ball bounce up and down for a length of time is motion cycling.

SYNERGY IN POST-PRODUCTION

Some systems can interface with other pieces of equipment to enhance the process of video animation. One of the most common interfaces provides VTR control, so the animator can record sequences without leaving the workstation.

Although some formal interfaces are available, no specific apparatus is usually necessary to combine the graphics system with the digital effects generator. This provides another approach for the artist who wants to create simple animation. At the least complex end of the realm of digital effects, loading an image from the graphics computer into an ADO (Ampex Digital Optics) or DVE (digital video effects unit) allows movement of the object without more than a single computer-generated frame. Over the picture of a clock, for example, we can key the image of a pendulum. Digital effects allow us to swing the pendulum back and forth without having to create multiple positions. The DVE performs these maneuvers by a process similar to tweening in the graphics computer.

Digital effects generators often allow the producer to manipulate two-dimensional images in what appears to be three-dimensional space. Rotation around x, y and z axes are possible, as are skew, perspective changes, and control of position and aspect ratio.

Frame Stores

As we progress from the simplest animation packages up the scale, other features and capabilities become more common. One is either an internal frame store or the ability to interface with an external frame store. Frame stores allow the artist to view animation in real time or near real time before dedicating the sequence to tape. Recall the frame buffer we described earlier, which is a computer memory space dedicated to holding one frame of video. The frame store is a device that can hold a number of frames, most commonly on magnetic hard disk or magnetic computer tape. Frame stores may hold, instead of one or two frames as in the frame buffer, 50 or 100 *seconds* of video. Rather than holding the frames in a space where a paint system can access them for creation or manipulation, frame stores are more storage oriented. Once the image is in the frame store, it can be recalled into another device for manipulation, or the frame can be used as an element in another image, but the computer image cannot be altered directly. The device can be used to receive output from a graphics system, recording the images as they are composed in real time onto a tape so that they will be available for further manipulation by other post-production equipment. The frame store can also be used as an input device, recording live video for frame by frame manipulation by the paint system, or to feed the images recorded from the paint system into other post-production equipment.

Digital effects generators with frame store combine some of these technologies. The digital effects generator or digital animation system possesses the capability to alter the image, and their digital disk recorders provide the images as well as recording the finished effects. "Layering" of effects has been a post-production technique for some time. Many effects may appear in a single sequence of video, and often these sequences are created one effect at a time. In the past, each effect would be put together and recorded on tape. That tape of the effect would then be played back as a source in another pass through the effects generator, and the next effect would be added, becoming another layer. The main problem with complex sequences requiring many layers was the generation loss experienced each time the partially constructed sequence was re-recorded with the next layer. Now, with products like the Quantel Harry, many passes can be made without going to tape. Since layers of effects can be produced entirely in the digital domain, unlimited passes can be performed without generation loss.

Integrating the simple graphics systems with other post-production equipment utilizes existing equipment to expand the capabilities of the paint system. If the director of the video department can allocate some funds toward the purchase of a new piece of equipment, and can afford a paint system, these techniques help stretch those dollars into as large a variety of new graphics capabilities as possible. If renting time on a graphics system is considered for a production, outside production charges may be kept to a minimum by bringing in some graphics produced on a relatively simple system for manipulation on equipment the producer may already have in-house. For example, a producer may create a company logo treatment and all the graphs and titles necessary for a program and carry them into the post-production facility to be enhanced with digital effects or to be dropped into a program with more sophisticated imagery. If an outside production facility is being hired, this may be the most cost-effective means of achieving an effect for the corporate client.

The Quantel Mirage

A dramatic example of the possibilities afforded by combining equipment is to manipulate your graphics system images in a Mirage. The Quantel Mirage is a $500,000 video image manipulator, which can twist and turn the television picture just about any way imaginable. It is not a cheap machine to purchase or rent time on, but it can quickly produce some spectacular effects. When the effect is defined and within the capabilities of such a device, buying an hour on the machine may make more sense than spending days trying to create similar sequences on a combination of less sophisticated machinery.

One of the standard functions of the Mirage, for example, is to take a two-dimensional video image and turn it into a three-dimensional shape, like a cylinder, cube or sphere. Take a grid on the screen and, in real time, transform it into a grid globe, which may spin or move around the screen. Such an effect is possible in a few minutes on the Mirage and would take far longer to construct and record on a paint system-VTR combination (if the artist is capable of producing the desired effect using a paint system). Most of the Mirage's capabilities do not come from predefined effects, but from the manner in which it may be programmed to distort a video image. The addition of one of the package effects, however, like the simulation of a turning page, may be just what the producer desires to enliven a program.

CONCLUSION

Joining the forces of the paint system, perhaps with animation capabilities, and other post-production equipment offers a far wider range of capabilities than relying on the graphics system alone. The combination allows you to get the most out of each dollar spent on the system, whether those dollars are spent purchasing equipment or on buying time on the system. Although the combined power of such system configurations can manipulate two-dimensional images in three-dimensional space and lend considerable visual interest to a production, they still cannot model images in more than two dimensions (except, of course, creating an animation of multiple perspective drawings in the cel style). The capabilities of three-dimensional rendering systems will be discussed in Chapter 5.

5 Three-Dimensional Imagery

None of the systems discussed thus far has been able to do more sophisticated graphics or animation than you could do in any other two-dimensional medium. The benefits of using computers in these applications include saved time, easier input into the video system and easier manipulation. What if, however, we want to record three-dimensional objects? We could enter them into a paint system in perspective renderings and record individual moves sequentially. The process is very difficult, perhaps even impossible, for virtually no one can reproduce perfect perspective. Thus, moving the objects about in space in this fashion would result in unnatural motion. It would also be terribly time-consuming, and therefore expensive. Customers requiring animated sequences involving three-dimensional objects would end up paying the artist for all the time necessary to copy 30 images per second of program and altering each one to create the illusion of motion.

The systems we will discuss in this chapter are vector-based systems. The objects are defined by wire frames; lines of specific length are arranged at fixed angles to one another. They exist as a series of coordinates and are thus independent of individual pixels on the screen. They can be manipulated mathematically without affecting other models in the computer universe. When these objects move on the screen, the computer needs only to render the moving objects and not the entire image. To create a fully rendered animation of smooth, shaded objects in real time, expensive frame store capabilities are still required, but vector-based systems are computationally more efficient and, thus, faster and more accurate.

To keep the technology affordable, almost all 3-D animation packages have offered a simplified version of the artist's construct in real time on screen before series are dedicated to tape. Most commonly, the test run utilizes a severely reduced palette, little or no lighting, and wire frames rather than finished surfaces to demonstrate the programmed motion. If the choreography of the elements of the picture behaves according to plan, the operator then requests the system to construct each finished image and record it to tape. Computers are used to communicate with a single-frame VTR onto which the completed sequence will be recorded. Commands input by the artist begin building the first frame of the series, lights are placed and textures wrapped, and the system lays the first image down on tape. The next frame is defined and similarly recorded. This assembly is a routine the machine can do by itself; many houses keep their systems occupied at night taking the day's animation and recording it after hours. The artist and the client can come in in the morning and take a look at the animated sequence.

Although some of the high-end machines we will discuss are unlikely to fit into corporate video equipment or production budgets at the present time, they may follow the pattern set by other computer systems and be affordable (perhaps even indispensable) tools in a few years. How many computers could be found on secretaries' desks more than a few years ago? How many secretaries could be as productive today without their word processors?

CONSTRUCTING MINIATURES AND MODELS

Animators have used miniatures and models for years, putting the recorded images through complex and costly matting and coloring steps to yield the finished image. Take, for example, the case of a metallic, spinning three-dimensional rendering of a corporate logo. A traditional film animation house may have handled it this way. Start with a three-dimensional wood model of the logo. Suspend it on dowels in front of a camera and film the desired move onto high-contrast motion picture film by moving the camera and spinning the logo on the dowels. Retouch each frame of the resulting film to remove the dowels suspending the logo. Make two negatives of this film and mask out the logo in one and the background in the other. Around the mask of the logo, expose a new film with the background color. Through the mask of the background, project the image of the spinning logo, and expose it onto film with the logo color. Matte together the resulting color films to assemble the final sequence. Sounds complicated? It is.

Constructing Mathematical Models

Luckily, graphic computers now have the capability of constructing mathematical models of three-dimensional objects and manipulating them in representational three-dimensional space. They have the capability of rendering shading and surfaces that appear so much like the actual objects that it is sometimes difficult to discern the difference between computer-generated and live video. Sometimes, the computer can create animated objects at less cost than constructing a set and taping the live scene. Of course, another major benefit of this capability is the construction of three-dimensional imaginary objects that do not exist in the real world.

CREATION OF THREE-DIMENSIONAL SPACE WITH THE COMPUTER

To create three-dimensional scenes with the computer requires, in the broadest sense, two steps: describe to the computer what the world looks like and tell it how you want to view that world. We will spend some time on the former process; the ease of the latter is the real power of the 3-D systems.

There are a number of ways by which we might go about describing the universe or, more important, the object at hand to the system. First, we may describe the object of our attention by means of a mathematical model. This is, as you may imagine, of limited utility to a video producer. Few of us pay attention to the mathematical relationships that describe everyday objects. This method is employed occasionally in the rendering of abstract concepts for instruction or simulation, like the drawing of a planet's orbit in a science video.

Rotation and Extrusion

Simpler and somewhat more useful for us is the ability to take a two-dimensional object and rotate it. These "surfaces of revolution" facilitate the construction of three-dimensional geometric shapes; a circle revolves to become a sphere, a rectangle revolves into a cylinder, and so on.

Related to this technique is extrusion, whereby a two-dimensional image is pulled out of the plane on which it is drawn. The flat image provides the template and is given depth by the computer. Extrusion can best be understood by considering the simplest extruder—the Play-Doh Fun Factory. A child selects a die, slips it into the toy and places a lump of Play-Doh under a large lever. Pressing down on the lever squirts the Doh through the die and creates a three-dimensional object whose profile matches the shape on the die. When we extrude an image on the computer, we create the "die" with the stylus and have the computer elongate it just as we elongate the die in the Fun Factory by pushing Play-Doh through it. Again, extrusion is useful but limited.

Solid Modeling

Next in sophistication is solid modeling. Images are constructed from a finite set of geometric shapes. These simple shapes are stored in the computer's memory and are used as building blocks to construct larger models. There are a number of applications for this technique within corporate video and broadcast, but its limitation is its inability to render free-form curves and images. The technique is especially popular in computer-aided engineering (CAE), because the geometric shapes are easily translated into machine movements.

Inputting Three-Dimensional Objects

Objects can be described by defining different sides with a digitizing tablet and puck or stylus as were used in two-dimensional drawing. Designating endpoints of sides and facets and communicating how this plane relates to others already input enables the computer to assemble the pieces into a three-dimensional whole. However, digitizing devices that record in three dimensions are more practical. The object to be input is placed on a tablet, and a stylus or mechanical arm is traced around the surface of the object. These tablets recognize not only where on the two-dimensional surface the stylus is, but how high off the surface it is. This permits the direct input of a three-dimensional surface.

Regardless of the means of input, the computer stores the model as a series of coordinates in a three-dimensional ''world.'' Objects in this world are defined and located according to their positions along three axes; x, y and z. Every point on the model has unique coordinates in the ''world coordinate system'' in which the object exists.

DISPLAYING THREE DIMENSIONS

Once the model has been constructed, we must instruct the computer on how we want to view that world. The computer must then portray it on a two-dimensional screen or paper. This is an entirely different type of operation than the one used to construct the model. It must take into account not only what is to be viewed, but the relationship between the viewer and the world. The appearance of the world will change depending on the angle at which it is viewed and the distance from the viewer. This introduces the problems of perspective, hidden line removal and shading.

The Viewing Plane

Translating a three-dimensional construct onto a flat screen is a process of projecting the world coordinate system onto a selected viewing plane, creating what is called a viewing pyramid. The viewer is at the peak of the pyramid, and from the viewer the four corners of the pyramid are projected. The viewing plane is a two-dimensional surface between the viewer and the model, perpendicular to the viewer and forming the base of the pyramid. The imaginary corners of the pyramid continue beyond the plane to establish boundaries on what can be seen by the observer from this point of view. It is analogous to the glass in the viewfinder of a camera. Its relationship to the viewer never changes, but its relationship to the object changes with the viewing point. An imaginary line between the viewer and the model is called a projector, and the intersection of the projector with the viewing plane is the location of the point projected from the model. For computational simplicity, only the endpoints and vertices of the model are projected, and new connecting line segments are produced on the viewplane.

Perspective

To give the model dimensionality, the computer most often uses perspective projection to map the model onto the screen. The simple principles of perspective were developed by Renaissance artists. Basically, to give the appearance of reality, distant objects must appear smaller than ones in the foreground, an effect called foreshortening. Also, different aspects of scenes appear brighter or dimmer depending on their proximity to the viewer and to a light source. Perspective projection incorporates both of these effects. To accomplish this, all projectors must converge at the viewer.

Normalization

The process of converting the view of a three-dimensional object onto a two-dimensional surface is a process of solid geometry called *normalization*. The normal is a projector, which runs from the viewer, the center of projection, to each endpoint on the model, running perpendicular to the viewing plane. Objects to the side of the viewing screen require normals that are slanted away from the center of the screen. The more slanted the normal, the more foreshortening required to represent that surface on a two-dimensional screen. At some point, as the normal gets more and more slanted, the foreshortening becomes so great that that part of the object vanishes from view.

Figure 5.1: Normalization.

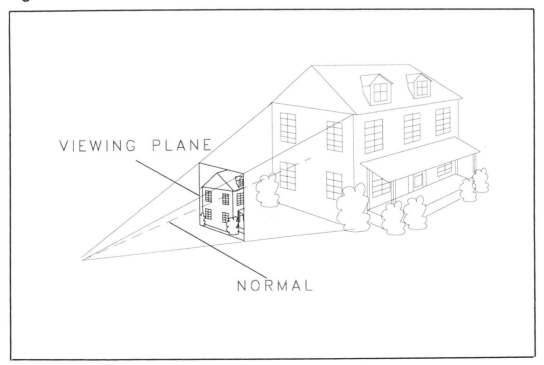

Clipping the Image

Once normalized, all elements of the translated image lying outside the viewing window are removed, or clipped. (See Figure 5.1.) The remaining points that are viewable through the window are run through the remainder of image processing. The rest of the model still exists, but is not used for the image being prepared for the video screen. This translation results in a new set of coordinates in normalized-device-coordinate space. Each point on the model within the clipping boundaries now has a corresponding point in this space, which will be mapped onto the video screen.

CONVERTING THE THREE-DIMENSIONAL MODEL TO A TWO-DIMENSIONAL IMAGE

The methods whereby the computer converts the three-dimensional model to a two-dimensional image are virtually the same from one computer system to another. The operator defines his viewpoint and distance from the model, and the vertices of the model are translated to endpoints on the viewplane, which are then reconnected to form a two-dimensional image of the solid.

Filling in the Surfaces

This construction and translation results in a view of the "skeleton" of the object. (See Figure 5.2.) Its surfaces have been positioned, but the model is still transparent, rendered as if a wire frame of the object in question. The viewer has the image of the model in perspective, but can still see right through it. We have constructed the imaginary steel framework, which defines the building, but our picture does not yet look like the building because we can see the structure of the far wall as well as the near.

Hidden Line Removal

In a solid object, only those surfaces that face the viewer should be visible; all other parts of the object should be hidden from the field of vision. (See Figure 5.3.) The most common method of hidden line removal was developed by Larry Robbins at the Massachusetts Institute of Technology in 1963. In its most basic form, this method compares all points with the same x and y coordinates, and deletes all but those with the lowest z values (corresponding to those points closest to the viewer).

Polygon Rendering

With vector displays, only wire frame images are possible in real time. This is because the vector system considers each point as separate and cannot distinguish any shapes that the line segments form. By generating graphics in raster-scan format, however, the computer can recognize areas between line segments as graphic elements (polygons), which may be colored, shaded or patterned. This polygon rendering is the most widely used method of turning a vector display into a more accessible raster-based image.

Figure 5.2: Wire frame. (See Plate 18.)

Courtesy John Stribiak and Thom Papanek

Figure 5.3: Solid Object. (See Plate 19.)

Courtesy John Stribiak and Thom Papanek

The use of polygons is simple when the image being rendered is uncomplicated. Many geometric shapes are polygons themselves and are easy to reproduce. Curved surfaces present some difficulties. The computer approximates a curve by creating a series of straight line segments along its perimeter, and simulates a curved surface by creating a patchwork of small polygons with straight line segments as edges. (See Figure 5.4.) The more complicated the surface, the more computation necessary to generate the image. The fuselage of an aircraft is an example of a very complex surface that requires many polygons to approximate. The computer creates a polygon mesh, a series of polygons with shared vertices and edges. These meshes are generated in the world coordinate model by assigning a specific number of polygons to a given area.

Hidden Surface Removal

There are two common forms of hidden surface removal: scan-line algorithms and the z-buffer approach.

Figure 5.4: Polygon mesh.

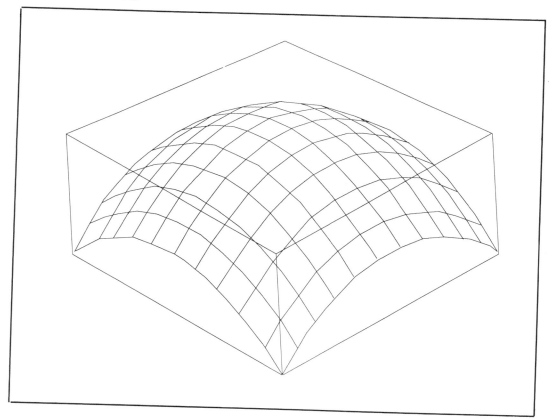

Scan-Line Conversion Algorithms

Once a model has been described as a polygonal mesh, it can be transformed to two dimensions following the same principles discussed for hidden line removal, but in this case, we are concerned with hidden surface removal.

Scan-line conversion algorithms pass all polygons defining an object and pass them through a "rasterizing" routine, which converts the surface of each polygon into a list of lines in three-dimensional space. The computer compares these scan lines from each polygon, sometimes called *spans*, to determine which lie closest to the viewer. Spans covered by other spans are eliminated, since the closer polygon would shield the farther from view. Only the remaining spans, the ones closest to the viewer, are passed on for rendering into the finished image.

The Z-Buffer Approach

In the z-buffer system, a section of computer memory keeps track of z values for all pixels on the screen. The computer then compares the z values for all pixels with the same x and y values and discards all but the lowest z values. These would be the values for the pixels that would be closest to the viewer. The result is a two-dimensional display of a solid object whose surface is composed of numerous, sharp-edged polygons. (See Figure 5.4.)

Catmull's Bicubic Patches

A method used to produce a smoother representation of surfaces was the subject of a PhD thesis by Edwin Catmull at the University of Utah in 1974. (The University of Utah has been significant in the development of computer graphics technology.) Catmull proposed that surfaces not be represented by a series of polygonal patches, but as "patches" that were the direct result of algebraic equations. His "bicubic patches" form parallelograms with curved sides and are defined by parametric equations with three variables, x, y and z. The variables correspond to the coordinates of the patch, and each is a function of two variables u and v. The sides of the patch are functions of u and v.

Second Order Derivative Equations

The bicubic patch was a breakthrough for computer graphics because it can be "squashed" into a perfectly smooth surface, thus ending the approximation of curves with polygons. The bicubic patch model has been improved upon from the standpoint of computational efficiency. A recent development has been the use of second order derivative equations to describe bicubic surfaces, which allows multiple patches to be described in a single equation. This technique has provided a solution to a problem in shading. When patches or polygons are used individually in shading, there are instances, especially in specular highlights (highly reflective, mirror-like surfaces),

Figure 5.5: Smooth shading.

Courtesy Artronics Corporation

Figure 5.6: Reflectance. (See Plate 20.)

Courtesy Wavefront Technologies

when adjoining patches differed slightly in color or intensity. The ability to treat more than one patch at a time, which Catmull's original model could not do, provides for smoother shading.

SURFACE MODELING TECHNIQUES

Highlights and Shadows

Assuming our model is not constructed of bicubic patches, we would now have a polygonal representation of a three-dimensional object, which can be colored as graphic elements. The next step is shading, which is important because it gives clues to perspective. The brightness of an object is a function of both the angle at which the light strikes it and the angle at which it is viewed. The computer can determine the brightness of an object by multiplying the intensity of the light source by a coefficient derived from Lambert's cosine law, a law of physics that describes the intensity of light reflected as a function of view angle. This shading method, however, produced polygons of uniform shade. (See Figure 5.5.)

Gouraud Shading

An improved method is Gouraud shading, developed in 1971 at the University of Utah by Henri Gouraud. Rather than treating each polygon individually, Gouraud treated the shading of an entire surface. He made a smooth transition of shading from one side of a polygon to the other, and therefore smoothed shading across polygons. However, the surfaces rendered with these techniques had uniform reflectance; there were no highlights.

Specular Reflections

Phong Bui-Tuong, working at the University of Utah in 1975, developed a method of displaying mirror-like specular reflections. His work was essentially an extension of Lambert's law. In addition to diffusing a calculated, uniform amount of light off a surface, Phong created a procedure that calculated how much light would be reflected off the surface back toward the viewer.

Like other methods of shading, Phong's procedure dictates that the sharper the angle between the light source and the viewing plane, the less light is reflected. Unlike diffuse reflection, though, specular reflection is concentrated around the normal. Instead of producing a uniform, matte surface as in previous shading methods, spectral surfaces would show concentrations of reflectivity where light sources would reflect directly to the viewer's eye. (See Figure 5.6.)

Phong's model is probably the most widely used lighting technique in computer graphics. It can, with a conservative level of computing power, render realistic surfaces. It still has shortcomings, though. Light sources are still assumed to be infinitely distant, so they cannot be included in the scene, and the model does not

Figure 5.7: Texture mapping.

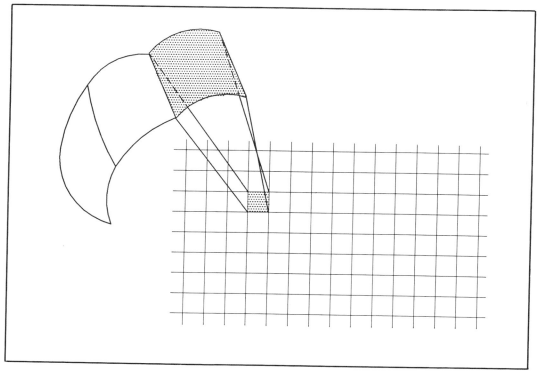

take into account true reflectivity, where light from one object bounces off another. In the real world, we see surfaces that reflect light from both light sources and other reflecting surfaces. Phong's model, like the ones before it, could show reflection from surfaces *only* from light sources, not from other reflecting surfaces.

Texture Mapping

A more sophisticated method of surface modeling, called texture mapping, was developed by Jim Blinn and Martin Newell, again at the University of Utah, in 1976. Their discovery was an outgrowth of Catmull's work in bicubic patches. The technique is based on the use of the four coordinate points, which define each bicubic patch as the four normals for each pixel on the screen. (In the previous techniques, like Phong's, only one normal per polygon or patch was used.) The result is the ability to display the precise three-dimensional orientation of each pixel. If the desired effect were a bumpy surface, an equation would be applied across a surface, which would alter, or ''perturbate,'' the normals of each pixel on the surface. When the model was subsequently lit, the pattern of the equation would appear, giving the illusion of texture.

Any texture can be created this way. Textures could be painted on the screen and subsequently mapped onto any surface. Using a similar process, two-dimensional patterns could be wrapped around three-dimensional objects. As the image is mapped

Figures 5.8a and b: Texture mapping applications. (See Plate 21 and 22.)

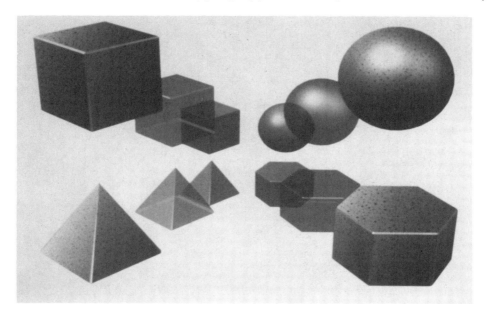

Courtesy Cubicomp Corporation

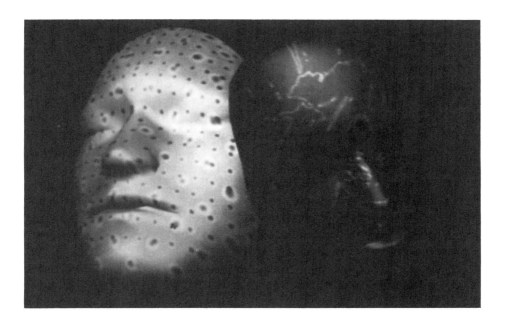

Courtesy Ian Jaffrey

Figure 5.9: Ray Tracing. (See Plate 23.)

Courtesy Robert Abel and Associates

onto a curved surface, the technique distorts the pattern as the normals slant away from the viewer, creating a "skin" which perfectly obeys the laws of perspective. (See Figure 5.7, 5.8a and 5.8b.)

Ray Tracing

Texture mapping is useful for those surfaces that are opaque and not reflective. Ray tracing is a technique that processes the characteristics of these complex surfaces into a finished image and can simulate highly reflective surfaces like chrome, refractive materials like glass and water, and transparent but colored materials. Ray tracing is a computer simulation of geometric optics. First, the computer is fed data about the characteristics of the surfaces in question. Such data would include, for example, the relectivity of the surface, its attitude in three-dimensional space, its color, and so on. The algorithm then starts at a pixel on the screen and looks into the picture. When the viewing line strikes an object, the characteristics of the surface it strikes determine what happens to the color of the pixel where the ray originated. If the surface is a flat color, that color is transmitted to the pixel. If it is a shiny surface, the proximity to the light source will determine how far from the color of the surface the pixel will be, to help account for highlights. If the surface is glass, the ray splits into two, one bending and continuing as it is transmitted through the material, the other reflecting off it and heading in the direction of the light source. The pixel becomes a blend of the colors the various rays tell it to be. (See Figure 5.9.)

Ray tracing yields highly realistic surfaces, but at the cost of demanding lots of computing power. Most ray tracing is set up on a mini- or supermicrocomputer and transferred to a much larger one for processing. This is the process of *front-ending*, which utilizes the graphics front-end/rendering engine equipment configuration described in Chapter 3. Images generated this way may even require a Cray (the largest and fastest brand of computer in the world, made by Cray Research) to use many minutes, even hours, to construct one frame.

Alternative Techniques for Lighting Computer Models

Because time on these very large computers is not widely available to producers of modest means, more common techniques of lighting computer models exist. Scan-line algorithms take into consideration the position of one or more light sources within the computer model and decide whether individual pixels should be in shadow or not, as the electron beam tracks its way down the television screen. There are substantial shortcomings to these methods; they cannot, for example, portray reflections, but are far more affordable, given the constraints of the average corporate production.

FRACTAL GEOMETRY: MODELING NATURE

When modeling physical, especially man-made, objects, wire frames and polygonal meshes often work quite well. Modeling nature, however, presents some new difficulties. Objects in nature are far too complex to be accurately represented by a reasonable number of line segments, and the resultant image would consist of far too many polygons for economical storage in the computer. Various attempts have been made to grow images from a limited number of commands, which act as rules to govern the growth of an object and instruct the computer to use a finite number of control points in order to draw an image.

Mandelbrot's Model of Self-Similarity

One method is based on a mathematical model developed by Benoit Mandelbrot in 1975 at IBM's Thomas J. Watson Research Center. Mandelbrot observed that many objects in nature exhibit the characteristic of self-similarity. That is, they are uniformly uneven at any magnification. A photograph of a beach from space shows it to be rough and uneven. When standing on a mountain and looking down at the same beach, it appears to be uneven to the same degree. In fact, the appearance of the beach is similarly uneven right down to the grains of sand of which it is composed. Mandelbrot believes that self-similar curves provide basic tools for understanding a wide variety of natural phenomena.

Fractal Surfaces

Mandelbrot's geometric model is a mathematical system, which utilizes a class of shapes that exist not in the conventional Euclidean space of lines, planes and solids, but in fractional dimensions between them. The coastline would exist not in one or two dimensions, but somewhere in between. According to this "fractal geometry," there are an infinite number of curves whose dimensions are between one and two. For example, the coastline might best be represented by a curve whose dimension is 1.7. Another curve with a dimension of 1.4, would best portray the leaves on a tree or the blood vessels in a lung. Fractal surfaces have dimension between two and three and can describe surfaces in nature with surprising accuracy. Depending on the dimension, mountains, clouds or foliage can be described.*

A Fractal Construct of a Mountain

Let us walk through a fractal construct and generate a mountain. Three parameters are input to the computer for a ''Mandelbrot set,'' the series of formulae that the computer will use to generate our model. The first parameter is length. It will describe the height of the peak. The second parameter is called the seed, which is a choice among a finite number of growth patterns, describing the degree of randomness that will be applied to the outline of our mountain. The third parameter is dimension, which, for surfaces, may be anywhere between two and three. Dimensions close to two will result in a hill with bumps on it; those closer to three will produce a jagged mountain.

Polygonal Subdivision

One method utilizing these parameters is polygonal subdivision. The computer builds a series of polygons according to the formulae into which the above variables have been inserted. The system works something like this. Start with a scene in which all surfaces are, for example, triangles. Divide each into a series of smaller triangles by joining the midpoints of all the sides and offset these points by a random amount to produce an uneven surface. Each new triangle can be similarly subdivided until the longest line segment resulting from a further subdivision would be shorter than some preselected minimum. Other polygons can be divided in the same fashion for objects of different shapes. This method enables the computer modeler to construct a fantastically complex surface, like a landscape, with very little input.

Mandelbrot's Impact on Computer Graphics

It is unlikely Mandelbrot realized the enormous impact that his technique would have on computer graphics when he developed this branch of mathematics. Richard Voss, also of IBM's Thomas J. Watson Research Center, was one of the first to use the technique to generate images and has produced perhaps the most impressive fractal renderings yet, including his Fractal Planetrise, which is a view of an imaginary planet as seen from one of its moons. The graphics world took a keen interest in fractals when Loren Carpenter, while at Boeing, produced an experimental film, *Vol Libre*. Carpenter created it for a flight simulator program in order to put the viewer behind the controls of an imaginary glider on a trip through a range of fractal mountains.

---

*Benoit B. Mandelbrot, *The Fractal Geometry of Nature* (New York: W. H. Freeman & Co., 1982).

Fractals, like wire framing, are merely a method of describing the superstructure of the object in question. The result is still transparent, and surface and illumination information must be added to the computer model to render realistic images.

"GROWING" IMAGES AND PARTICLE SYSTEMS

The Growth Algorithm

There is at least one other major method of developing three-dimensional images within the computer. The scientists researching these techniques hypothesized that, while attempting to simulate the look of nature in computer graphics, perhaps it would be reasonable to mimic the process of nature, to grow images rather than build them. This line of thinking led to the Growth Algorithm by Yoichiro Kawaguchi — a system that could render images of natural forms like flowers and seashells.

Attempting to recreate nature at first created an interesting mix of simulation and illusion. Distant hills and mountains do not look as crisp as close ones, because mist in the atmosphere obscures objects at a distance. In order to reproduce this optical effect scientifically, Jim Blinn, now working at NASA's Jet Propulsion Laboratory, built statistical models of clouds filled with randomly distributed spherical particles. The thicker the cloud, the more dense the distribution of particles, and the greater the chance that a ray of light passing through the cloud would strike a particle, be scattered and not reach the viewer. Simultaneously, some light entering from the viewer's side of the cloud would strike a particle and be reflected back.

While scientifically accurate for the process of light transmission through mist, this technique did not render a very realistic cloud and presented some difficulties with haze and fog. Again, the question arose whether describing nature statistically was necessary, or could nature's processes be imitated.

Particle Systems

A fascinating process called Particle Systems was developed by Bill Reeves at Lucasfilm. The technique is based on a system composed of random particles with two parameters: the length of time particles will live and the direction in which particles will tend to move. The system itself is given parameters: density and frequency of new particle generation. Within these parameters, particles behave according to a stochastic (random) model. The system comes close to simulating Brownian motion, one description of the way particles move in nature. Tracing the motion of a particle results in an element in the final picture, a short line segment or curve. The record of an entire system has resulted in the surprisingly realistic rendering of nebulous objects in nature, like clouds, or other groups of objects, which are of no particular form, like clumps of grass. In the cloud, millions of infinitesimal particle traces become an amorphous mass; in a field, each particle trace is a blade of grass.

Other than nebulous objects, these systems are being used to model scenes in which the viewer needs only an impression rather than details, like a landscape obscured by haze. This is a fascinating new direction for graphics research, modeling not the objects, but how they are perceived by the eye. Film animator, Skip Battaglia, pursued the same conceptual line when he used blurs of color rather than discrete objects for grass blowing in the wind and flying insects in his film *As the Frog's Eye Sees*.

Particle systems made their debut rather dramatically in the film, *Star Trek II: The Wrath of Khan*. The detonation of the Genesis Bomb produced a wall of fire, throwing off sparks and gradually spreading across the surface of a planet. The fire was a complex particle system, and each spark thrown off was a miniature system.

GRAFTAL GEOMETRY—A COMBINATION OF TECHNIQUES

Graftal geometry was developed by Alvy Ray Smith, also of Lucasfilm, in 1983. The technique is neither fractals nor particle system, but includes characteristics of both. The major distinguishing feature of using graftals is the absence of the need for computationally expensive random numbers. The process involves recursively dividing basic forms according to a basic rule with limited subdivisions. Although appearing to have infinitely increasing detail, as could be achieved with fractals, there is actually only enough to give the impression of this degree of detail.

As fractals best represent mountains and rocks, graftals are best demonstrated with trees. A single trunk divides into branches, these into smaller branches, and on down to the twigs. The resulting data strings describe the form of the object being portrayed, and the computer converts them into images. Since only three or four subdivisions of the main element of the object are being performed, the object is far simpler to represent this way than with fractals.

TRANSLATING THE IMAGE TO TAPE

Once the world has been described to the computer, and the machine has been told how to take a look at that universe, the view or sequence must be transferred to video. In the case of still imagery, the method is fairly simple: call up the image and use it as a feed in post-production. Of course, if we were interested in a still image, it would most often make sense to render it on a paint system and avoid the complexities of trying to describe a three-dimensional object to the computer. Introducing motion, however, presents some of its own problems. The most important of these is the computing power required to store more than one frame for rapid access and display.

The two-dimensional systems discussed in the previous chapters were primarily paint systems. These systems, by definition, define their objects by color. Images are constructed with a series of pixels, which are of a specified hue, and their arrangement on the screen defines the picture. The paint system is bit-map based because the object on the screen is nothing more than a collection of colored squares arranged in a specific fashion. To create motion on such a machine requires drawing a new image for each frame in which motion occurs. Thus, to move objects through manipulation of a paint system is computationally time consuming.

IMPACT ON THE VIDEO PRODUCER

The techniques and equipment applied to the generation of an image have a multitude of ramifications for the corporate producer. The most obvious and important is how the selection of image technology will affect the price of the sequence. The more advanced the technique and the more difficult the image to construct, the more expensive the technology required to produce it. As noted in Chapter 4, there are times when the use of more advanced machines may actually save money, due to the compressed speed at which the effect can be accomplished. A useful generalization, however, is that the more complicated an image and the more complex the technique to produce the image, the more resources will have to be allocated. Chapter 6 will discuss costs and budgets further.

Another way in which a selected technique affects the producer is the level of sophistication required of the production team. This applies principally to the corporate facility manager contemplating the purchase of equipment; the producer hiring services can usually discuss the effect in terms of desired outcome. The prospective equipment purchaser, however, must face the fact that the systems with greater capabilities require more technical expertise than do the simpler systems. For paint systems, the artist need only learn a few simple computer commands, disk care, and how to power the system up or down. When contemplating more complex three-dimensional or animation systems, the operator will be faced with the possibility of picking up a working knowledge of a high-level computer application language in order to construct and manipulate images.

Plate No. 1

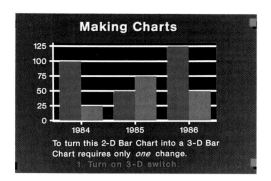

Plate No. 2

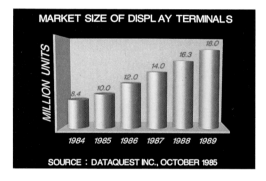

Plate No. 3

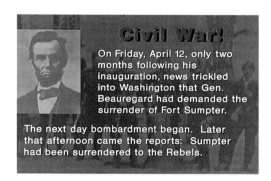

Plate No. 4

Plate No. 5

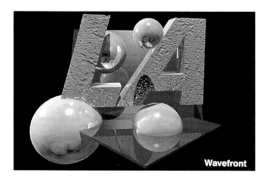

Plate No. 6

Plate No. 7

Plate No. 8

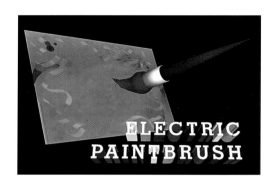

Plate No. 9

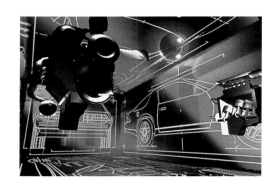

Plate No. 10

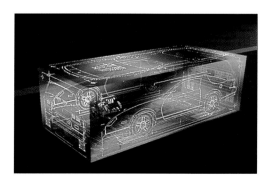

Plate No. 11

Plate No. 12

Plate No. 13

Plate No. 14

Plate No. 15

Plate No. 16

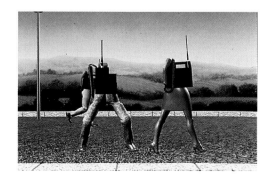

Plate No. 17

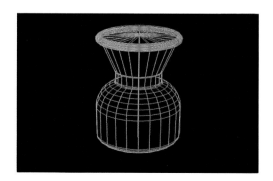

Plate No. 18

Plate No. 19

Plate No. 20

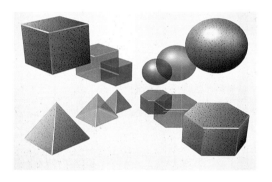

Plate No. 21

Plate No. 22

Plate No. 23

Plate No. 24

6 Time and Money in Production

Thus far we have discussed some of the history of computer graphics, a few of its applications in video and how computers generate images. Of course, very little of this information would have much value if the cost were unaffordable. The issues of cost and the time required to assemble a graphics sequence are central to the decision to use computer graphics. Unfortunately, it is often difficult to provide specifics to guide the producer who has never before used computer graphics. Anyone who has been in the position of quoting prices on a video production knows how complicated a bid can be. Everything affecting the finished program exacts a cost, and each factor must be considered in the estimate. Pricing graphics is difficult.

This chapter will help shed some light on the factors influencing the cost of producing finished computer graphics for video. The principal factors in determining production time will be addressed, and four sample systems will be examined. Chapter 7 will take up the topic of evaluating the purchase of equipment.

The cost factors involved in generating computer graphics are very complex. Comparing prices will not give an indication as to what is difficult within the environment of the computer and what a computer will do almost automatically. Compounding this problem is the state of affairs of the computing world, generally. By the time this text reaches the bookshelves, it is entirely possible that a new system will have been announced that performs graphic functions at a fraction of the cost of the same process today. In addition, equipment dealers will charge different prices according to the service they later provide, and manufacturers and independent producers discount prices depending on the quantity of equipment or service required by the customer. It is important to understand that this information may not be absolutely infallible. Absorb whatever you consider important from this book for use as guidelines, not as absolute rules or price lists.

For these reasons, much of this chapter will be presented in the form of case studies. We will speak about general capabilities of a number of systems with "ballpark" prices, outline some of the factors to be considered when pricing graphics production, and review several specific projects that have been created in the corporate realm.

THE PRICE OF GRAPHICS

First, we will discuss the major factors involved in the cost of computer graphics. The main elements of graphics costs are, generally, overhead, equipment, personnel and post-production.

Overhead

Overhead is an expense most often accounted for only in commercial production houses. It is not uncommon for a corporate video department to have to earn its budget every year in charge-back income, but rarely does that department need to pay for the space it occupies. These costs must, however, be included in the calculation of an hourly rate, whether it be for billing outside the company or for an internal charge-back. There may be monthly expenses for security, accounting, advertising and so on. These costs must all be factored into the charges that reach the customer. We will not elaborate on how these are calculated, but mentioning their existence is necessary. When preparing budget requests for management, factoring in the overhead demonstrates an understanding of the overall costs to the corporation and can demonstrate how a new piece of equipment is profitable within the context of the firm, not just the context of the department.

Equipment Expenses

The next factor involved in the cost of graphics is equipment expenses. The cost of a piece of video equipment must be recovered over its useful life. Sometimes, there is an additional percentage to cover financing charges resulting from the equipment's purchase or lease. Two main points that differentiate computer graphics equipment from most other video equipment are the useful life and the software.

Useful Life

Keep in mind that the life of a piece of computer graphics equipment is substantially shorter than the life of most other video equipment. How long do you expect a camera to be in service? Four or five years, perhaps? Longer with a retubing? How about a waveform monitor? Waveforms in museums are sometimes newer than some still in service. That's fine. They still do their job and do it well. Computer imagery equipment, on the other hand, is charged with the role of providing contemporary visuals to an audience with high expectations for modern appearance and technical quality. For the last several years, the rate of change in technical capabilities of computer graphics systems has been accelerating. The more sophisticated animation becomes, the shorter the useful life of the equipment involved, and the less time that a machine on the "leading edge" will stay there. The manager of a video department housing computer graphics equipment must amortize his hardware as fast as possible or risk being caught with a house full of obsolete machinery, machinery that can no longer perform its job because of technical inferiority.

Software

Another aspect of graphics equipment cost that video producers have not traditionally had to account for is programming. Software is the soul of the computer graphics system and can account for half of the cost of obtaining and installing a system. The capabilities of the system are defined by the software. Like digital video effects equipment, a skilled operator can produce images that stretch the boundaries of the system or utilize the equipment's capabilities to maximize the value of a production, but the software ultimately provides the limits on what can be accomplished. Most video departments cannot afford or justify a resident programmer for their graphics computers, so the software provided by the manufacturer delineates the facility's capabilities.

The upshot of this is that, because the equipment is software-driven, it will be more expensive to run over time. Software advances arrive faster than hardware improvements. Thus, the purchaser of computerized systems will face either paying for software upgrades or switching to competing systems. If available software does not improve quickly enough, the useful life of the system is shortened.

We do not mean to say that everyone employing computer graphics turns over equipment every two years. Image West of Studio City, CA, continues to employ a Scanimate (manufactured by Computer Image Corporation), one of the first analog computer graphics tools. They generate imaginative and effective graphics with it. However, Image West uses some of the most recent graphics and effects technology as well. Talented editors, directors and artists will always find ways of embracing many levels of technology to communicate. As the technology progresses, however, expectations of how a program should "look" will evolve, and this evolution implies technological upgrading in the production suite. It is the same challenge faced by producers of private television, as when nonbroadcast video witnessed the introduction of color or digital effects. Keep in mind that the manager must amortize his equipment inventory so that today's system is not relied upon exclusively too far into the future.

Staffing

The next factor influencing the cost of graphics is the personnel involved. Graphics production is often more than a single-person enterprise. For simpler productions, one person may generate the graphic design, execute the graphics and insert them into the program. As the production gets more sophisticated, however, other people may be called upon to generate storyboards, construct computer models for animation or even assemble the elements into a final sequence. The situation is the same as when a location shoot becomes more complicated; the requirements grow from a videographer to include a gaffer, tape operator, audio technician and grips.

Post-Production Time

Finally, remember that in most productions, computer graphics necessitate additional post-production time. Once the graphics have been created, their assembly, motion and inclusion in the program often take place in the edit suite. The producer calculating a graphics budget must account for not only the equipment and personnel involved in the graphics suite, but also for the editors, tape machines, switchers and effects generators used to merge the graphics with the other elements of the production.

PRODUCTION TIME

Next we turn to the factors involved in the length of time it takes to produce a graphics sequence. They are scripting/storyboarding, keyframing or model construction, rendering and scheduling.

Planning a graphics sequence can range in complexity from a brief description of a still image or a table of numbers to a complete storyboard with detailed illustrations of each major change in the sequence. At its most basic, preparation for a graphic may consist of handing a table of figures to the artist with instructions to make it into a bar graph. At its most sophisticated, detailed drawings may be necessary from which a model will be constructed in the computer.

Simple treatments as a substitute for storyboards, as described by Chuck Althoff in the case study on the Thermal Magnetic Duplicator, which follows, may take a day or several days (at eight hours per day) to produce. Complex storyboards or model illustrations, as were necessary in the construction of the robot in ''Brilliance,'' described in that production's case study, may take several weeks.

Key Frames

Once the preliminary design has been approved, the next step is the generation of the key frames or models around which the sequence will be built. In the case of graphs or titles, this is the finished product. In animation, the key frames are the video version of a storyboard. Some people, in fact, may substitute these for storyboards. A video image is created for each major scene change. The point of this step is to create the minimum number of images necessary to display all the major design elements and scenes so that everyone may see enough of a sequence to approve of its appearance. Once the production of animation is under way, any changes required by the client will extend the production time for the graphics and increase the cost, or make alterations impossible without reconstructing the entire sequence.

Rendering

Once the keyframes have been approved, the images must be rendered. Rendering time varies widely with the level of graphic complexity. Still frames or simple motion may be achieved in real time and recorded directly to tape. Color cycling, two-dimensional motion or graphs, which are built on the screen, can often be produced in a number of hours once the keyframes have been defined. Complex surfaces, motion in three dimensions or multiple passes through effects equipment require more time to produce.

In one of the cases that follows, a mouse runs a maze and later appears eating his cheese reward at the end. The chewing mouse took a number of hours to animate and record. In the motorcycle liability case study, creating technically correct motion of a motorcycle moving over a grid took between three and five days. Rendering highly reflective surfaces, such as the chrome robot of "Brilliance," may take up to 20 minutes *per frame*, and this only after the models and key frames have been generated. The fact that the generation of sequential images may require a substantial amount of production time reinforces the importance of including the graphic artists early in the planning of the program.

Scheduling

Scheduling presents the final determinant in estimating the time required for graphics production. With adequate preparation, resources can be scheduled as needed in order to provide minimum interference in the smooth flow of the production process. Keep in mind that there are usually fewer graphic artists and systems than there are videographers and cameras. If the system must assemble animation or render complex scenes, it introduces a relatively inflexible element into the time requirements of production. Allow enough time in the production schedule to provide for access to the appropriate equipment, or allow enough money to contract out the process to a production house, and perhaps even allow enough money in your budget to pay for inserting your rush job into the production house's schedule.

FOUR SAMPLE SYSTEMS

Brief descriptions of four sample systems will give you some idea of what capabilities are available at different price levels. These vignettes should not be considered endorsements, and the prices of these systems may have changed between the publication of this book and your decision to hire graphics services or purchase the equipment yourself.

Inovion PGS III

The PGS III professional graphics system is a stand-alone paint system, based in a proprietary personal computer with a Motorola 68010 microprocessor. Its paint capabilities include airbrush, geometric shape construction by rubber-banding, image grabbing from film or video, copy and sheer modes, expand and reduce modes, three levels of zoom and high resolution. It is capable of interfacing with virtually anything for input or output. Output is possible to video or film. The system comes with a mouse graphic input, and the system can be used with a bit pad. Standard system configuration includes a 3-inch floppy disk and an internal hard disk. The price of the PGS III is $7295. For additional information contact Inovion, 250 E. Gentile, Layton, UT 84041, (801-546-2850).

Artronics VGS — Video Graphics System

The Video Graphics System is a foundation graphics system available at low cost with multiple upgrade options. The system may be purchased complete or in individual modules. The modules are 8-bit paint with AniMagic real-time animation, 24-bit paint with flash-grab and software for true 2-D cel animation with optional VTR control, and a 3-D Model Shop and Animator.

The 8-bit paint module offers user-definable brushes, text generation and a palette of over 16 million colors — 256 colors are displayable at any time. AniMagic offers unlimited simultaneous color cycles and fades, each with distinct start and stop points and independent speeds. Real-time multiplane animation is also possible.

The 24-bit paint module allows simultaneous display of 250,000 colors. Colors can be mixed as with paint on a traditional palette and can be rendered with degrees of transparency. This module offers instantaneous flash-grab, and the captured images can be retouched and combined into cel-animation sequences. Commands can be entered from the bit pad or keyboard. A dual frame-buffer system allows simultaneous work on two images and offers a Save/Restore function.

The 3-D animation module allows the creation of models in which the artist can define geometry, texture and the appearance of each object. Object creation techniques include extrusion, surfaces of revolution, fractals and 3-D text. Once created, the 3-D animator allows the artist to move and change the attributes of objects, lights and camera angle. Features include translation, rotation, scaling, acceleration/deceleration, camera pan, tilt, dolly and zoom. Rendering capabilities include wire frames, flat shading, Gouraud and Phong shading with anti-aliasing, 16 light sources, specular reflection, transparency, texture mapping and hidden surface removal with interpenetrating objects.

The system is based in an IBM PC AT-type computer with an internal 20 megabyte hard disk drive and 1.2 megabyte floppy disk drive, $512 \times 512 \times 32$-bit image memory, Genlock, RGB output. Input is through a bit pad via puck or light pen. Options include digital film recorder output, single frame and VTR controllers, 40 megabyte removable hard disk. The system costs less than $30,000 with one module and includes delivery, one day's training at the customer's facility and two additional days training at Artronics. Additional modules cost between $7000 and $8000.

For further information contact Artronics, 300 Corporate Court, PO Box 408, South Plainfield, NJ 07080 (201-756-6868).

Cubicomp Picturemaker/60 R

The Picturemaker/60R is Cubicomp's standard video animation system. Three-dimensional modeling software includes extrusion, surfaces of revolution, cutting, drilling, surface warping, cross-sectional modeling, helices and spirals, mirroring, beveling and metamorphosis. Models can be constructed of wire frames with or without hidden surface removal. Up to five independent light sources can be defined and shaded. Images can be anti-aliased, smooth-shaded, made translucent, Phong-shaded or texture-mapped.

The system arrives with eight outline text fonts with more available as options and the ability to generate user-defined fonts. Titles can be scaled, italicized, kerned, extruded or beveled.

Animation is accomplished with motion scripting in 3-D space using key frames, and software offers storyboard display, velocity specification, computer-generated shot sheets and automated output to video or film.

Paint functions include anti-aliased airbrushes and custom brushes, patterns and graded fills, lines and shapes, blurs, color mixing, mattes and masks, moves, rotations, zooms and 2-D type fonts. The system also offers an Undo feature.

The system is an 80386-based modeling and animation system, which comes with a Compaq Deskpro 386 computer equipped with a 80387 math coprocessor, with 40 megabyte hard disk and 1.2 megabyte floppy disk and 4 megabyte expansion RAM. The system includes a digital tablet with puck and stylus and 19-inch RGB monitor, and a CS-16 frame buffer with Genlock with two image frames. Also included is the 60R Race board (rendering accelerator computer engine), which makes rendering of finished images an average of 20 times faster than a similar computer without the board.

The Picturemaker/30 comes as a turnkey system for less than $69,500. The price includes two days of training. For further information contact Cubicomp Corporation, 21325 Cabot Blvd, Hayward, CA 94545 (415-887-1300).

Alias Research Alias/2

The Alias/2 3-D graphics design system is a turnkey workstation, which offers computerized design in real time and realistic 3-D animation. The artist accesses the system through the User Interface, a set of programs using on-screen artificial intelligence. Three-dimensional space is accessed through simultaneous windows of front elevation, side elevation, top view and perspective view. Any change in one view is simultaneously reflected in all other windows.

Animation of unlimited objects, lights, surfaces and textures is produced along motion paths rather than traditional point-to-point motion. Sequences can be quickly previewed in wire frame. Animation is controlled using variable on-screen timing curves, automatically generating images between key frames.

Lights can be animated, unlimited in number, color, diffusion and specularity. There is complete control over ambient light color and brightness. Animated surfaces and backgrounds can be created with control over color, specularity, diffusion, refractive index and transparency. Surface shading can be simple Lambert, Phong or Blinn shading for metallic effects. Surfaces can be bump-mapped for texture effects.

There is control over camera position, movement, focus, focal length and field of vision as well as eye viewing position.

Using a complex object modeling system, modeling can be accomplished using only a few unrelated data points. Traditional methods required thousands of unrelated data points for even simple objects. Objects can be created, transformed and scaled easily. Scene Assembly groups, copies and links separate objects and groups of objects.

Paint packages are available to be used in conjunction with this system for between $2000 and $3000. The system is capable of rendering to a resolution of 1024 $\times$ 1280 lines.

A quick rendering function allows previewing of still images and animated sequences using simple shading.

Text functions include control of character kerning, justification, word spacing and unlimited point sizing, fully rendered text recall and a large selection of fonts.

The graphics front end is a Silicon Graphics Workstation, based on the MIPS RISC chip. It is equipped with 32 megabytes of internal RAM. Peripheral storage is accomplished on an online Winchester hard disk capable of storing up to 2.2 gigabytes. Peripheral optical disk storage can also be fitted to the system.

The system is capable of interfacing with a wide variety of input and output devices and offers a number of options to increase memory space and speed.

The turnkey system starts at $65,000, and users receive three days training either at Alias or their own facility, with ongoing training available.

For further information call Alias Research Incorporated, 212 Carnegie Center, Suite 202, Princeton, NJ 08540 (609-987-8686).

CASE STUDIES

The following five case studies present a more complete picture of various aspects of computer graphics. In each, we present the communications problem to be solved, how the producer settled on computer graphics to solve it and some of the details involved in completing the job. These examples, although far from exhaustive, represent a range of communicative and technical complexity.

Case #1: Motorcycle Liability Demonstration

In 1985, a liability suit was brought against a motorcycle manufacturer, alleging that it had failed to install a piece of commonly available safety equipment — a metal bar that extends from the sides of the motorcycle and is designed to prevent injuries to the lower legs and feet of the driver if the bike lands on its side in an accident. The litigant, who had purchased one of these bikes, was in an accident and claimed that the safety device would have prevented his injuries. The manufacturer, like most, did not believe the device would provide this protection without cost; they maintained that the bar could stop the vehicle suddenly once parallel with the ground, catapulting the rider over and free of the bike. The resulting free flight of the driver, they believed, could cause more injuries than those sustained in an accident without the leg protectors.

The defense team in the case was faced with the responsibility of communicating the manufacturer's opinion of the effects of the crash bar to the court. They needed to persuade the court that they were not negligent and that their decision not to install the bars was based on safety considerations. The company decided that a description of the situation and the physics involved would not make the point clearly enough. Replicating the accident with real motorcycles and real people would be destructive, dangerous and impossible to record in adequate detail. They turned to simulation in the form of computer animation.

Ken Creasman of Yale Video was assigned to the project. His solution was to build a motorcycle in the computer, hold the image in the middle of the video screen and indicate motion along the road with grid marks beneath the bike.

Using what is now the Cubicomp Picturemaker 30 (Cubicomp Corporation), Yale took about a week to construct the model of the motorcycle in the computer. Constructing a three-dimensional model of the vehicle generated costs to the client of between $5000 and $6000. So much for the easy part. The critical task was to produce a motion sequence that would accurately portray the bike moving at specified speeds and demonstrate the likely outcome of a crash.

Creasman points out that generating the motion was the challenging part of the production. First, he describes the system they used as not being the easiest to move. It is not quite the same as pointing a finger down the road and saying to a stunt man,

Figure 6.1: Stages of a motorcycle accident.

Courtesy Yale Video

"Go that way at thirty miles per hour." More important, the grid, which was to move beneath the motorcycle, had to move at a rate proportionate to the size of the bike to represent lines at measured intervals on the road. When the accident occurred and the motorcycle fell on its side, the movement of the grid had to slow down and stop at the same relative rate as an actual motorcycle would come to a stop in the same circumstances. (See Figure 6.1.) If the lines, for example, represented intervals of 50 feet, the motorcycle had to move from one line to the next within a specific amount of time. To deviate from this rate would render the sequence inaccurate, and it would not provide the visual argument the company was trying to make.

The execution of the motion path, accomplishing the move of the grid underneath the bike, took from 24 to 40 hours. A great deal of time was spent setting experimental speeds within the computer and watching and timing wire frame tests of the sequence. These tests were necessary because the computer does not accept motion commands in miles per hour. The team at Yale took the wire frame construct of the motorcycle and rider and tried a number of speeds before the image travelled at the rate specified by the manufacturer's engineers.

The entire sequence took approximately 40 hours to accomplish, but that time was spread over several weeks because they had a number of jobs to accomplish within the same period. The total running time was 20 seconds. The total cost to the client was between $7000 and $8000.

How well did the video perform? Creasman reports that the case never went to court. If the animation enabled the manufacturer to avoid a trial, the cost was more than justified in saved legal fees. What did the sequence provide? According to Ken Creasman, "The video made a statement to the other side of the case. It may not have proved the case, since it was never used as evidence. But I can't help but believe that having this video made the people bringing the suit say, 'Hey—these guys really have their act together.' I'm sure that helped keep it out of court."

Case 2: Mouse in the Maze

Mike Conklin, a video producer at Eastman Kodak Company in Rochester, NY, was faced with a familiar problem, one he was accustomed to solving. This time, with less than one week to prepare, he was assigned to record a seminar on motivation and to transform it into a video production that could be distributed to managers who could not attend the talk.

After receiving a copy of the text of the talk, he set to work on defining the contents of the presentation. Taping even an interesting presentation and failing to embellish it would not hold the audience's attention. That would be especially inappropriate for this production; the speaker was a widely recognized consultant on motivation. Since much of the content of the talk was quantitative, he turned to computer imagery to generate graphs and tables to help illustrate some of the salient points. But the production needed something else. It was still too static.

In one segment of the presentation the speaker referred to reward systems for employees. He compared the reward systems to teaching a mouse to run a maze, a fairly standard procedure in experimental psychology. That was the example Mike Conklin needed. He could create an animated mouse striving to conquer the obstacle course and finally being rewarded with a mountain of cheese at the finish line.

He contacted Tanya Weinberger, then staff animator for Telesis Productions, Inc., and described his needs. They met to design rough storyboards and discuss the goal and the length of the piece. She generated rough keyframes and a short sample of the mouse running part of the course to demonstrate how it would look. "What Tanya created was very close to what I wanted," Conklin said. "I told her to add a pink belly and pink up the cheeks a little; not to make it a rodent, but to make it a character." She had these samples ready for Conklin two days after they originally met.

Conklin came back a few days later for the edit session. When they began work on the mouse segment, there were three techniques involved. Before the session, the animator and editor laid down the opening of the segment in single frames on 1-inch tape. The mouse appeared at the beginning of the maze, turned his head, moved his tail and ran inside. This section was joined to an overhead view of the maze. The mouse, viewed from above, moved his legs frantically through a continuously repeated paint animation sequence. He was made to move around the maze by keying his image onto the maze and moving him with ADO. When he finished the course, the shot was moved back to his eye level, to show a now-fattened mouse leaning against a much smaller mound of cheese, chewing happily. The mouth and whiskers of the mouse moved through color cycling, and the motion was recorded in real time using the paint system as a feed. They hooked the paint system to the switcher directly and ran the effect in real time as the VTR rolled. This contrasts with capturing the image in a still-store or as a still on tape, in which case the still-store or VTR would have been the feed.

The sequence lasted two minutes. There were almost 20 minutes of graphics in the 30 minute production; the balance was the tabular and graphic information mentioned before. The mouse sequence was the most expensive and time-consuming of the graphics involved. From first consultation to final editing took approximately 40 hours over a 5 day period. The audience loved it. The graphics lent visual appeal to the program, and the mouse sequence invariably made the viewers chuckle.

Case 3: Thermal Magnetic Duplicator

Corporate producers are often faced with the challenge of developing a visual presentation of something that cannot possibly be photographed live. If the subject is purely conceptual, as in the previous case of the Mouse in the Maze, the graphics treatment can be equally conceptual. Such visual metaphors have great communicative value.

But what if the producer has to depict an object or process that cannot be seen directly, either due to its enormous or minute size, or because the process involved occurs in some inaccessible place? What if the process is invisible to the human eye or camera? And what if the depiction has to be of a level of sophistication and accuracy to please even the most scrupulous engineer? This was the tough assignment faced by Chuck Althoff, Production Supervisor, TV Services, at the DuPont Company.

As in most large companies with production departments, Chuck's studio acts as a vendor to other departments. This vendor-client relationship is carried right through the billing process. When approached by a "client" to produce a job, TV Services provides them with all the services they require, either in-house or by contracting with an outside vendor for production, graphic or editing facilities. The ultimate responsibility for the project's budget remains with the initial clients; they decide how much money to commit to their project and, based on advice from Chuck's people, what production techniques or effects are appropriate.

When Althoff was asked to produce the initial product marketing tape for DuPont's new high-speed video duplication process, known as the Thermal Magnetic Duplication System (TMD), his first job was to learn how the system works. "It incorporates a laser," says Chuck, "and it's a contact printing system, where it's impossible to see the laser heating the tape and what magnetic particles are doing under the heat of the laser."

Not even the laser beam could be seen. Unlike the familiar colored, glowing beams, this variety of laser is invisible to the eye and the camera. Chuck knew immediately that computer graphics were the only means of illustrating the system.

Because of his background in film animation, Chuck is never surprised at the high cost of computer graphics. In fact, he claims to be amazed at how quickly and inexpensively most graphics can now be produced. Many of his first-time clients do not share that opinion, but Althoff sees that changing rapidly.

"I tell them that it costs, and the cost is dear," he says, and he lets them know up front that they will probably have to go out-of-house for computer animation. (Although TV Services has some computer graphics equipment and plans to add more in order to increase productivity, he feels the majority of computer graphics production will continue to be out-of-house.) Althoff reports that as his clients have become more familiar with the use of computer graphics and have experienced positive results, they have become much more accepting of the costs.

In the case of the TMD projects, Chuck prepared a treatment and presented it to his clients to help them determine what the production would accomplish and whether the expenses were justified. His department uses such treatments rather than full-blown storyboards, which are expensive and time-consuming to produce. He saves storyboarding for the high-end, big-budget projects, many of which are done in conjunction with outside agencies. With a treatment, he shows his clients dubs of previous programs that incorporate computer graphics to demonstrate their effectiveness and to give a comparative cost.

Once agreement was reached on the type of graphics to be used and a budget, the TMD project was sent to an outside production facility to create the computer graphics. A simple motion graphic was produced, using a combination of a Dubner graphics system and digital effects. In the final edit, a live close-up of the TMD machine's drum head dissolved into a graphic of the same component, rendered in the same scale. The motion graphic then showed the movement of the original and duplicate videotapes along the machine's tape path and the laser heating the tapes. It also depicted the polarization of molecules within the tapes, demonstrating the actual (but invisible) cause of the transfer.

Like most producers, Althoff looks back on the project with an eye toward how it could have been improved. He had been reluctant to use the particular Dubner system, feeling that it produced graphics with a two-dimensional quality not totally appropriate to the application. However, he found it the best way to achieve the clients' objective within budget. The total cost of the project was approximately $8000. About half that went toward the computer graphics and digital effects.

The clients were extremely pleased with the results. Besides being technically correct, the graphics added life to the program. The graphics allowed the viewer to observe a process that is normally impossible to see.

Thanks to Althoff, the clients received an added benefit. Chuck was able to create a computer graphic company logo for use in this and future programs for only $2000. Better still — from the clients' point of view — the logo could be paid for from a different budget. The clients were impressed and liked the logo so much they decided to use it as the actual product logo on the machine.

Case 4: Water Crisis

How does a water crisis in southern California lead to imaginative new uses of computer graphics?

When the amount of available water from the Colorado River was drastically reduced, the Metropolitan Water District, a consortium of several cities in southern California, had to look elsewhere to augment their region's water supply. The logical place to look was the northern part of the state, where water was plentiful. But when the voters defeated a proposition that would have allowed the consortium to construct a canal to carry water to the south, the Metropolitan Water District had a dilemma on its hands. How could it convince the voters of the critical need for a project to transport this vital resource from the north?

Thus, a full-scale public information campaign was launched. Included in this campaign were plans to educate the general public, school-age children and politicians with the facts surrounding this complex ecological issue.

Before long, Image West, a production facility in Studio City, CA, became a focal point for this production effort. Originally, they had been approached by Brad Laven, an independent producer working for the Metropolitan Water District. This first contact had been to arrange for some animation needed for the project, but as often happens in the production business, those initial discussions soon led to something entirely different.

"We started negotiating not only to do the animation, but to actually become the studio of record," said Image West's producer, Carl Reichman. And that is just what happened. Image West provided Brad Lavin with an office, and he continued to head the project from under their roof, with immediate access to their facilities and expertise. Soon Lavin had a full staff working on the project and a formal contract with Image West.

The project then developed on several fronts. Plans were made for shooting the live footage that would be needed, and researchers scoured the Metropolitan Water District's archives for stock footage that could be used. Storyboards were developed for the animation. This was to include kinescope animation as well as 3-D wire frame images.

"One of the first major pieces of animation was the opening," says Reichman. He described a concept involving a graphic view of the earth as viewed from space, and a "point of view" journey, zooming down through the atmosphere, flying over California's central valley, soaring low over an expanse of water and finally over a dam. It was not actually a 3-D graphic, but could be described more accurately as a special effect involving both live video and animated graphics.

The next element was a fully rendered piece from a 3-D wire frame. "A map of the United States stands up," says Reichman, "and we see the whole country. The map is a wire frame, and it falls down and lays flat. As it's pitching away from us, it's our point of view that we're beginning to fly over the United States—over the wire frame. As we fly over and approach the west coast, California extrudes from the map and the rest of the wire frame falls away. We then fly over California and see all the state in three dimensions." The final 3-D image is a major graphic element used frequently to illustrate such things as the water table, evaporation, distribution into the soil and pathways of the different water systems.

Illustrating these complex geographic and ecological systems proved very challenging. It had to be done in a way that gave viewers an understanding of the "big picture" and motivated them to make some important choices for themselves and their communities. Only graphics and animation provided a means of showing interrelated systems, which, although surrounding us, are too large to observe in their entirety. By executing these graphics artfully, the motivation to act was also provided.

The project was approached just like a feature film. Everything was scripted and storyboarded, complete with scene numbers and timing. But like a feature, a certain amount of flexibility was allowed. In fact, the team produced what they refer to as a "dirty dupe," a rough edit on black and white film, assembling all the available stock footage in order to see where the holes would be. This told them what additional stock footage they would need, as well as what animation and live production was required.

When Image West was first approached in the fall of 1986, they were told that the Water District wanted to spend roughly $20,000 on the animation for the project. As the job grew, so did the budget. After December 1986, when the contract was signed for Image West to become the studio of record, the budget for the entire project was set at $82,500. The animation portion of that budget grew to between $25,000 and $30,000, reflecting the increased commitment of everyone involved.

Two versions of the program were completed in June, 1987—about six months after the original contract was signed. These included a full-length twenty-minute version and a six-minute re-edit for showing in schools. "About 25% of the full-length version is graphics and animation," says producer Blackwell. "Many of the same visuals will be used in the short version, but it's a whole new script and really a separate production."

According to Reichman, the kind of growth this project experienced is not unusual. "It happens a lot with us when people will come to us to do one thing and they see the range of what we're able to do. They say 'Can you do this and this and this?' and it usually turns out that we can." What may be less obvious, but equally true, is that the arrangement enabled Brad Laven, as executive producer, to deliver the production to the Metropolitan Water District at a significant savings. By having all production work originating from one full-service facility, a great deal of time was saved, and much costly coordination between separate facilities was eliminated. And it is nice to note that an important public service was rendered as well.

Case 5: Brilliance

One of the principal reasons there are applications for computer animation in corporate video is that all media producers are charged with capturing and holding an audience's attention. If the materials used in the educational setting are not memorable, their capacity to teach disappears. This requirement of instructional video is exaggerated in the production of advertising. Think of a broadcast commercial as a lesson. The average advertising producer has sixty seconds or less to teach (or, better yet, convince) the viewers that this deodorant will make them more attractive, that only one particular toothpaste will make them kissable, and so on. Furthermore, the advertiser does not have the opportunity to sit the audience down specifically to absorb this material, and those people who do view the presentation sometimes start class with an ''incorrect'' set of beliefs.

We will thus conclude our studies of computer-generated graphics projects with one constructed for this most demanding ''educational'' setting. This example also offers a glimpse of what lies behind a very sophisticated commercially produced animation project and demonstrates what goes into producing the highest-end animation.

In late 1984, Ketchum Advertising of San Francisco had the assignment to present a utilitarian but ordinary product in a fashion that would stimulate new interest in its use and offer it some glamour. The client was the Canned Food Information Council, and the product was the aluminum can. What can a producer do to generate new interest in a product that most people take for granted, has been around for years and offers little aesthetic or technological appeal? Robert Abel and Associates' solution was to develop a new technology to present the message.

The finished product was a thirty-second commercial in which a chrome robot, comfortably leaning back in the easy chair in her futuristic ''space condo,'' described how, even in the year 3000, she recognized the brilliance of storing food in such a durable, inexpensive and efficient container. (See Figures 6.2 and 6.3.) Two things in particular set this animation apart from what preceded it: the life-like motion of the star of the commercial, and the rendering of her chrome skin.

The robot in the spot moves with elegantly human character, not with the mechanical or marionette-like quality that animated forms frequently exhibit. One of the clever reasons for this is director Randy Roberts' plan to use a live model as the basis for the robot's movements. She was filmed performing the moves that were choreographed for the commercial. On the film her joints were marked with dots, which were then traced by a computer. Around these dots, the body and limbs of the robot were constructed to form a vector graphic.

Another significant factor in achieving lifelike motion was a computer program called *Direct*, which Kim Shelley developed for Robert Abel and Associates. *Direct* dictates and controls the motion of computer models hierarchically, building ''trees'' along which motion is transferred. Motion at the root of an object translates into motion on the limbs. Says Richard Baily, who modeled the robot in the computer, ''Because Kim's program ties the parts of the model together, it imposes useful limitations on the maximum range of motion of each body part. The thigh and calf, to take a real-life example, are limited by the fact that you can bend them, but can't really twist them. *Direct* provides a very realistic framework.''

Figure 6.2: Sexy Robot.

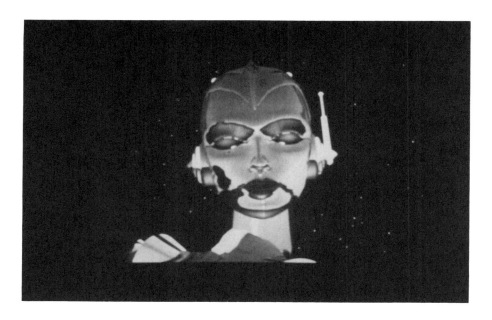

Courtesy Robert Abel and Associates

Figure 6.3: Android spokesperson relaxing in her space condo. (See Plate 24.)

Courtesy Robert Abel and Associates

"With this breakthrough," says director Roberts, "we could achieve motion that appeared organic instead of mechanical. From the first motion tests, it was clear that we'd captured remarkably lifelike gestures."

Advances in software also enabled the Abel team to render the chrome skin of the robot. The faceted surfaces of the robot's body and limbs, constructed of millions of polygons, were rounded and shaded to the point where they appeared completely smooth. Roberts commented shortly after production, "The fact that we were able to make shiny, smooth surfaces, with clear, accurate reflections, is an indication of how our software has really come together."

Combining advances in their rendering software with Shelley's *Direct* program, the creative team was able to not only provide lighting to the model, but to include lights in the scene as objects, which could be made to move and cast light in specific directions with specific patterns of light. Roberts reports that this development "allowed us to light a scene as if it were being shot on a stage, which gave us the chance to try some dramatic effects."

An Evans & Sutherland vector graphics system was used for modeling the robot and her environment; a Gould model 6080 for some of the processing; and, for the final rendering, the IRIS system by Silicon Graphics. The finished images were output directly onto 1-inch videotape. Creative Director, Robert Abel, said, "With scenes output directly on 1-inch video, avoiding all film transfers, we were capable of executing in a 12-week period what would have been impossible just a few months ago."

Production also eventually utilized quite a bit of the staff at Abel, in part because of the developmental nature of the technology being applied to the spot. Kim Shelley reports, "There's a lot of interaction between the people who develop the software and those who use it. Almost every technical director at the studio worked on 'Brilliance,' plus four of our full-time software people and our part-time people." The credits list fifteen Abel staff members.

The dedication of talent and equipment paid off. The commercial eventually amassed half a dozen of the most prestigious awards offered in animation and advertising. More important, recall scores calculated for the commercial after only two broadcasts rivaled campaigns that had run for months. "Brilliance," however, was not inexpensive to create. The final cost was between $150,000 and $175,000.

Abel commented on the effort to produce the spot and the studio's relationship with the advertiser: "The client and the agency both demonstrated the kind of trust in us that makes you want to give them an award-winning piece. They recognized that we were attempting to take their concept to the limit, and they let us go for it. When they finally came to view what we'd done, their reaction summed up what we were all feeling: It was just the way we thought it would be . . . except better!"

7 Bringing Computer Graphics In-House

A corporation with a video facility may consider outfitting itself for the capability of producing computer graphics or animation. When the needs for computer graphics dictate and the budget permits, there are advantages to having the capability internally. Not long ago such an option was virtually inconceivable because of the cost of the hardware. Once exclusively in the realm of mainframes, the creation of broadcast-quality computer graphics has, over the past few years, come within reach of the capabilities of the personal computer. We will discuss some of the advantages and disadvantages of owning such a system within the video department and describe some of the characteristics of a PC-based system, which allows the generation of video graphics.

CONTROL

There are some distinct advantages to having the internal capability to augment video production with computer graphics. An obvious one is the control the corporation has over the time and priorities of the system. When contracting out-of-house, any particular graphics job is but one of many that the facility is responsible for delivering. Production houses will do their best to perform their services in the best interests of all their clients, but the fact is your schedule and deadlines have significant competition from other projects at a commercial facility. It is unrealistic to expect that your project should jump to the top of the schedule because your deadline is tight or you experienced delays in script approval. The facility will try to help you as much as possible and may be able to return projects in surprisingly short time, but to expect this consistently of a commercial facility is not realistic. With an internal system, you can delay all other projects to accommodate a rush job (although this should be an emergency procedure only).

Control of access to proprietary or classified information may also be a consideration. Some corporate videos involve topics of a sensitive nature. Government contractors may recognize this clearly, but corporate confidentiality is also a serious matter in most firms. When graphics are used to explain a sensitive topic within a video presentation and the segment is created by an outside firm, it may be impossible to assure confidentiality. That is a fact of life when uninvolved parties handle material that does not influence their security clearance or competitive position.

An internal system may offer more creative input and control than does a contract with another company. The ability to walk down the hall to see how graphics are taking shape or the access the artist has to the producer for questions during production can have a significant impact on how well the finished graphics conform to the original concept. The producer should not hover like a vulture behind the artist, of course, but proximity has its benefits.

COST EFFICIENCY

One obvious reason for incorporating computer graphics into the repertoire of an in-house facility is to save money. If the company produces enough video and already contracts out for a great deal of computer graphics in many of its productions, it can be a means of reducing production costs. All costs involved, however, need to be taken into consideration. As addressed in Chapter 6, overhead, staffing and additional equipment support must be factored into the equation when calculating potential cost savings.

As corporations become more sophisticated in their production capabilities and because they want to take advantage of the benefits of owning equipment, there is a new option open to the corporate producer. More thoroughly equipped corporate video departments are now offering their facilities for hire. They are soliciting clients outside the corporation as would any production company. Prices for these facilities are often very competitive. The added benefit for a corporate producer is the ability to deal with a facility that was designed to solve the same kinds of communications problems the producer is out to solve. Before opening their doors to the public, these departments had only internal clients and may very well have gone out on occasion to contract for production services. The industrial producer will be speaking with people of the same orientation.

EVALUATING THE NEED

When the possibility of acquiring an internal system exists, the importance of performing a needs assessment cannot be overemphasized. The principal elements of this study are volume of work, sophistication of the proposed system, relevance of the technology and the prospects for expansion. Does the corporate facility currently have a high enough volume of work to keep the proposed system busy? Are there enough productions under way or being planned that would incorporate enough computer graphics so that the investment would pay for itself? If possible, services on this system could be offered to producers or companies out-of-house, but this should not form the basis for a purchase decision.

Are computer graphics the appropriate solution for the applications envisioned? Be wary of purchasing an expensive graphics system for the sole purpose of adding bells and whistles to productions. Assure yourself that the computer is the best method of addressing the needs at hand before writing the purchase order.

What needs can be projected for a proposed system? Many production houses have elected not to invest in animation or three-dimensional systems because the technology is still progressing at an astounding pace. Understand that any system you purchase today may be inferior in just a few months. This does not apply as much to the lower end of the spectrum. New and improved systems will be introduced to the inexpensive range of the market, but it is unlikely that these improvements will render any of the current equipment obsolete. In the realm of the more sophisticated paint systems and the three-dimensional animation systems, which carry significantly higher price tags, the worry is justified.

If there is likely to be heavy or increasing usage of the proposed system in the foreseeable future, it might make sense to purchase equipment that surpasses the current requirements. The ability to grow into a system will effectively prolong its usefulness to the company. In fact, buying a step above the current need can help shield against the spectre of obsolescence. Owning something closer to state-of-the-art technology leaves you not as far behind when the leading edge moves farther out. Only when your capital equipment base is heavily involved in state-of-the-art or your clients' loyalties are dependent upon leading-edge effects do you need to worry about keeping pace with the most up-to-date technology.

FINDING THE APPROPRIATE SYSTEM

Consider using benchmarks as a tool for comparing systems. Develop or analyze one job with typical difficulty and complexity and compare how well different systems will perform. Compare systems specifications in relation to your most common needs. How much latitude will you have with each system? Will some of your most common projects test the capabilities of the machine? Will you need to be creative in your utilization of the system to complete a standard production, or will it only require straightforward operation of the functions?

Sample Different Systems

To get a sample of the advantages and pitfalls of different systems, rent time on a number of systems. Send the person from your staff who will be the principal operator of the in-house system to complete a few similar projects on different systems. Have your person watch, listen, ask questions and perhaps assist the operator of the system at the production house in order to get a "feel" for the different machines. This will also give you the opportunity to compare the time requirements for similar jobs on each system. This, in turn, will allow you to calculate the operating cost of each machine. This information will help in determining whether such a machine will be a profitable addition to your facility.

Demonstration Tapes

As you would watch a demo reel to evaluate a production house, look at a demonstration tape of each machine's capabilities during your assessment of different systems. Ask a representative of the manufacturer to sit you at the console of the workstation and show you or your artist how various functions are accomplished. How familiar with computers is the person who will be operating this system? Is this a system that requires software skills, or are the functions graphic and self-explanatory? Must the operator turn from his work to use the keyboard and the status screen often, or can most functions be accomplished on the graphics display screen with the stylus or mouse?

Training

How much training does the manufacturer offer? Do they have an access number for customers to call with procedural or operating problems? For some of the more sophisticated systems, customer support can mean the difference between a productive asset to the video department and a time-consuming technological problem child.

Flexibility

How flexible is the design of the system? Are there help functions in the software? Have there been upgrades offered in the past to keep the system current, and does the manufacturer contemplate any in the near future? How well is the system documented? One of the chief complaints about microcomputers is that manuals actually tell very little about the system and the likely problems. Other manuals may describe the operation of the machine in detail, but in a virtually unintelligible manner. Are the manuals offered with your graphics generator easily understood with a minimum of coaching?

Customer Evaluations

Ask the distributor for other customers whom you may call. They can offer firsthand evaluations of the operational difficulty of the system, its bugs, and how well it interfaces with the rest of the equipment in the suite. They can give you ideas about the capabilities of the machine when in use. They, too, were once in the position of looking at a variety of systems before deciding on this one and would be able to tell you why they made this selection. Compare their needs and applications to yours.

PC-BASED GRAPHICS SYSTEMS

Now that we have briefly explored the issues a video manager must resolve before considering the purchase of a system, let us take a look at the elements of a PC-based system. PC graphics systems are spotlighted because they are the systems that have brought computer graphics within reach of most corporate producers. Many of the commercially available computer graphics generators are based in personal computers (or at least microprocessor-based), which makes their hardware similar to that of the personal computer.

One advantage of purchasing a complete system is that the manufacturer has assembled all of the elements into a package, which will have been tested and (we assume) proven to function together. If there is a problem with the function of the system after purchase, there is one source to turn to for help or recourse. If the system is assembled piecemeal, chasing down technical difficulties is a far more complicated undertaking.

HARDWARE

The computer that really opened up the opportunity to create effective and professional-looking graphics on a relatively inexpensive machine was the IBM PC/AT. This product and its clones (computers that are constructed on the same architecture but manufactured by other firms) combined a powerful microprocessor (the 80286) with an open architecture that allowed other electronics companies to create specialized peripherals that interfaced simply and efficiently to achieve specialized applications like video graphics. These peripherals, in fact, have come to actually control the graphics production process, using the central processor of the PC only as an overall manager, which controls system functions, file operations, and so on.

Additional Memory

One alteration, which is necessary for adaptation of the PC for computer graphics, is the addition of memory. Temporary storage of the image in process often dictates the addition of two to four megabytes of memory. Mass storage is also required so the computer has somewhere to keep the image once it has been created. If, for example, one image occupies 500 kilobytes of space and is part of a ten-second animation segment, provision must be made for 150 megabytes of storage. This storage may be on hard disk, large-scale floppy disks, streaming tape or a combination of these.

Adaptation of the Processing Circuitry

Next comes adaptation of the processing circuitry. The central processor is usually complemented with an 80287 math coprocessor to speed the calculations involved in rendering an image. Specialized input and output ports must be provided to enable the computer to communicate with its new peripheral devices used in graphics: the digitizing tablet, the VTR, a video camera, and perhaps a digital film recorder. Most of these specialized ports come as part of the plug-in cards, which transform the PC into a graphics workstation.

These graphics boards must provide some important features or be capable of interfacing with devices to accomplish these adaptations. The system must be able to differentiate between a sufficient number of colors and communicate them to the VTR. The upper limit on most of these systems is 16.7 million colors. This may seem a bit more than any one image should require, and many systems are capable of displaying only a fraction of this number of colors simultaneously. Remember, though, that a large number of very similar colors is necessary to accomplish shading, highlighting, and anti-aliasing.

Communicating with Video Equipment

Next, the system must possess the capability of communicating with the rest of the video equipment involved in recording the graphics. Most computers capable of producing color graphics output their video as separate RGB signals. The graphics system must be capable of NTSC composite output, the facility must be equipped with an encoder, which converts the individual signals into a single composite signal, or the graphics must be fed into a post-production suite capable of handling component video.

The PC must be capable of generating RS-170A sync (the standard synchronization signal used in video equipment) or of accepting a sync source from outside. A useful characteristic is the ability to genlock to the external sync source. Without some means of genlocking the computer to the video system, the frame buffer would frequently be out of sync with the rest of the system. The encoder would constantly be trying to compensate for the difference between computer and house sync. The result may look fine on a composite monitor, but a VTR attempting to record this unstable sync signal would end up with an unacceptably poor quality recording. Sophisticated post-production equipment would have even more difficulty with the resultant recording, making further manipulation of the graphics impossible.

Some plug-in boards, like AT&T's Targa Series, offer RGB and composite output and input, allowing the computer to digitize an incoming video signal.

The VTR Controller

The brain of the computer graphics system is the VTR controller. Once a frame of video has been composed by the computer, the controller issues the proper commands to the VTR to enable it to record the image. Some plug-in boards, like the Diaquest DQ-422, can control two VTRs at once—useful when a graphic is to be dropped into a videotape sequence. The board uses one VTR as source and one as recorder and can be instructed to drop the graphics into the sequence at the appropriate edit point.

The Time Code Generator

One more piece of useful support equipment is the time code generator. Successful recording of the graphics and successful editing down the line require exact positioning of the images on tape. The presence of SMPTE time code makes further processing or manipulation of the program by other video equipment much smoother.

SOFTWARE

Paint Software

Two-dimensional paint software is generally self-contained. That is, there are not separate functions that must be worked in concert to produce an image. There are varying attributes, to be sure, and different features, such as frame grab, but the ability to paint is the basis of the software.

Animation Software

Animation software, on the other hand, must provide four primary functions: an object editor, a scene editor, a scriptor and a controller.

The Object Editor

The object editor is used to create three-dimensional constructs in the computer. The user can create flat objects and rotate, extrude or revolve them into three dimensions, or create objects one aspect at a time. These objects, built of polygonal meshes, become data files within the computer.

The Scene Editor

The scene editor takes the objects created with the object editor and adds attributes of color, shading, light sources and so on. Some scene editors are interactive in that they allow the user to view the object being enhanced as it is being worked on. The scene editor performs the functions of hidden line removal and anti-aliasing, according to one or more of the techniques described in Chapter 5.

The Scriptor

The scriptor allows the animator to describe the motion of the objects, which have been created in the object and scene editors. Often, this is the most difficult step in the process. Simple linear motion is easy enough, but more complex motion involves more sophisticated scripting, often with hierarchical structure. In describing a person walking, for example, the legs must move in relation to the torso, the knees must move in proper alignment with the hips, the feet in accordance with the knees, and the whole body must move forward in relation to the size and speed of the step the figure is taking.

The Previewer

One of the drawbacks to working on a small computer is the time involved in rendering a finished image. Coloring and lighting an object may take 10 to 20 minutes per frame. Complex scripting, however, is exceedingly difficult without some real-time feedback. The PC alternative is called a previewer. It usually presents the objects in wire frame, as described in Chapter 5, and real time so that motion may be observed before being committed to a sequence.

Background Creation

Once the objects and motion have been described, the backgrounds are created. A paint system is usually included in the animation package, and it is used to create backgrounds, texture maps, and so on. The creative use of the paint functions allows creation of effects in $2^{1}/_{2}$ dimensions. Technically, this is cel animation, but can be used to simulate three dimensions. Consider a sequence involving moving around city streets. Two-dimensional views of the city can be moved in front of each other, creating a convincing illusion of moving around the city without having to build a time-consuming and computationally burdensome model of a city.

The Controller Program, or Compositor

Once all of these elements are in place, the animated sequence may be created. The objects and backgrounds are composed and moved according to the motion script by the controller program, or compositor. The controller determines the placement of the objects within a specific frame of video and coordinates the recording of the finished frame of video onto tape.

THE DILEMMA

Now you have some orientation to the elements of a graphics or animation system. You may choose to assemble the components described above, or purchase a system already assembled by a manufacturer. Some systems with their components and capabilities were described at the end of Chapter 6.

What if we cannot justify the purchase of an entire system? What if the systems we can afford do not offer the level of sophistication our audience requires? What if our managers are not yet convinced of the importance of purchasing a dedicated system? There is, of course, an alternative. Computer graphics and animation production houses abound. Our knowledge of the systems will probably prove useful when looking at the equipment and capabilities of an outside production facility, but there are other considerations when contracting for outside services. We therefore turn our attention to hiring an outside production facility in Chapter 8.

8 Contracting Outside Services

When the video manager cannot yet justify an in-house system, cannot obtain the budget to acquire one, wants to learn more about graphics before committing to them, or does not feel the need for in-house capability, there are many options for finding commercial production services to accomplish a graphics objective.

REASONS FOR CONTRACTING OUT-OF-HOUSE

There are a number of persuasive reasons for going outside to find production services, especially the first few times a producer uses computer graphics in a video. The most obvious reason involves technological availability and experience. The more sophisticated the effect or sequence being considered, the more complex the instrumentation and the more expertise required of the production crew.

There is also the issue of lead time associated with the purchase of a computer graphics system. Once the machine is pulled from the box, it will take some time for a member of the video staff to develop enough familiarity with the system to begin generating graphics according to a reasonable production schedule. Production groups and facilities allow the producer to make use of experienced personnel until he gains a more thorough understanding of the medium, or until an in-house system can begin producing at a reasonable pace.

In addition, contracting outside services allows the producer to choose from a wide variety of equipment. Bringing capabilities in-house limits the options for the types of graphics that can be incorporated into your programs. Until the production team becomes familiar with the capabilities and idiosyncrasies of several systems, it would be difficult to make an informed purchase decision.

113

An outside production company may be able to bring a fresh, creative perspective to a job. When the producer or director has to worry about all aspects of a production, having an art director to design graphics may mean an increase in the efficiency of the production team. In addition, an outside graphics facility may have experience with creative solutions to similar communications problems and may offer an enlightened approach to your problem.

Facilities that are heavily involved in producing computer graphics offer the opportunity to generate graphics quickly and efficiently because of their breadth of experience in creating for a number of clients. In addition, some facilities offer packages or libraries of effects and graphics to customers, who do not possess the capabilities of producing them in-house. The collections may include digital effects, backgrounds, short animated sequences, and a host of other products created by their post-production and graphics equipment.

THE PRODUCTION SERVICE MARKET

The computer graphics and animation production market is broken down into several relatively distinct strata, defined by cost and sophistication. At the very top end are about half a dozen houses that produce custom work almost exclusively for the broadcast and theatrical markets. These companies produce the most striking and sophisticated imagery; companies that largely define state-of-the-art. The principal disadvantage is cost. Animation from this group costs several thousand dollars per finished second of video. It would not be too unusual to spend over $50,000 for an ID (a standard program opening or trailer, which usually includes a logo) at one of these facilities.

In the next lower price range we find video production and post-production companies that have developed computer graphics capability. Many of these firms are very well equipped with slick 2-D and 3-D equipment, and additional editing and DVE facilities with which to enhance their graphics. Companies that own a Mirage, Bosch FGS 4000, sophisticated paint systems, multiple frame store capabilities, ADOs, and so on, usually fall into this category. Buying time in these suites often costs from $400 to $750 per hour, but the wide capabilities of the equipment mean the money is often well spent, providing the budget allows. An attractive ID might cost between $10,000 and $25,000.

There are many firms that rely less on broadcast production and do more corporate and industrial work, and so have more affordable facilities. These companies have augmented their production or post-production equipment with two-dimensional or paint systems. As the price of three-dimensional and animation systems falls, more of these outfits are adding these facilities as well. An attractive ID or brief opening can often be produced at one of these firms for $5,000 to $10,000.

As computer graphics become more popular and more specialized, a new type of production house has come onto the scene. Rather than offering a full range of production or post-production services, this new type of facility produces only computer graphics and animation. The very top-end houses largely belong in this category, but there are more affordable facilities in this category as well. These firms often produce computer imagery for more than just video. Presentation graphics, slides, and other visuals come from these firms, but they are increasingly able to produce video graphics and animation.

The Design Group

Another option is the design group or, in some cases, the independent producer. These are companies that regularly produce graphic sequences for video production, but which do not own their own equipment. They may be computer artists, or may be video or film producers who have become specialists in computer-based visuals.

It might appear that, if the design groups are using an outside production facility, you, as the client, might be better off (at least financially) going directly to the firm that owns the computers and employs the artists. You could employ the operator and the system directly and thereby eliminate the profit the designer collects on the transaction. Two factors must be considered in this decision, though. First, as discussed above, it is important to have someone who can fill the role of art director. A familiarity with the functions of the equipment as well as experience in design for video are necessary. A producer without these credentials, who approaches the computer artist directly, may fall short of achieving the communication goal.

Second, and more important, there are distinct advantages to operating in an atmosphere free of equipment. Any facility, which owns machinery, must employ it as often as possible or lose money. In addition to the money spent on an equipment purchase, the resources invested in the equipment include its operator, the electricity it draws, and the floor space that could be allocated to potentially more profitable equipment. Worse, an artist or designer who spends all day creating images with the same equipment tends to think in terms of what that system will produce. Thus, an artist, who constantly uses a specific animation system, may not realize that your problem can be solved more efficiently (or as easily at lower cost) by employing a paint system with, perhaps, an ADO move (a special effect, as created by the Ampex Digital Optics effects generator). A designer who has working relationships with several production companies does not run as much risk of falling into a rigid approach.

CHOOSING THE RIGHT PRODUCTION COMPANY

In looking for a service to produce your graphics, begin with your peers. Many corporate producers have experience with outside production companies and can give you names, numbers and evaluations. The International Television Association (ITVA, 136 Sherman Ave., Berkeley Heights, NJ) publishes a list of members, and you can consult this for members in your area who may have contracted out-of-house. *The Video Register and Teleconferencing Resources Directory*, published by Knowledge Industry Publications, Inc., White Plains, NY, lists production sources, including production companies and corporations using video. Finally, many of the video trade publications contain annual directories, which are broken down geographically and by services offered.

The Demo Reel

When attempting to decide on a firm to contract for graphics services, there are a number of helpful evaluation techniques. One obvious and useful resource is the demo reel. Every production company has one (or several) to show prospective customers what the company is capable of producing. For veteran producers, this is often a valuable way to compare houses. However, if the client does not have experience evaluating a facility by its reel, especially in computer animation, this tool can be very misleading.

Evaluating the Demo Reel

Since the reel is a sales tool meant to demonstrate the limit of the facility's capabilities, it can lead to what Susan Davis of Varitel Video calls the "sexy robot misconception," making reference to an award-winning commercial by Robert Abel and Associates. The difficulty arises when the client mistakenly assumes that any effect or sequence on the tape is affordable. Looking at a collection of the most impressive imagery a company has to offer without keeping budget constraints in mind is of limited usefulness to a producer.

When evaluating a demo reel also keep in mind that the client might have been responsible for a lot of the creative work. Many production companies are hired to execute effects and animation based on the creative work of the client's art director. That this work should end up on the demo reel is not a deception; the facility did produce the work. If, as the client, you are in the market for creative as well as technical services, it pays to know how much of the demonstration reel originated with the production house's art directors and artists. Ask questions and make notes on specific sequences from the tape. Find out who worked on various effects and graphics as artist, editor and director. Can you talk with those people before committing to the company? Will they be available to work on your project?

Evaluating a Production Company

One of the best ways to evaluate a potential subcontractor is to call current customers. People who have worked with the company can give a good indication of the strengths and weaknesses of the facility and its staff. If some of the references are producers who work with a number of facilities, they will be even more valuable because they can offer comparisons with other companies. Of course, the production house will only refer you to satisfied clients, but these contacts still provide a wealth of information. Happy customers are in a better position to offer an objective evaluation than is the salesperson or representative of the facility.

Soliciting Bids

The definitive method of comparing companies, at least from a budgetary point of view, is the solicitation of bids on a specific job. Comparing bids is far superior to a comparison of rate cards. Many firms shy away from putting emphasis on rate sheets and some even prefer not to offer them for examination. There are too many variables involved in a video production (much less a piece of animation!) to assume that two facilities would use comparable lengths of time and amounts of resources on the same job. That would imply that all have similar equipment, that there are no differences in maintenance levels, and most important, that all production houses are staffed by personnel of identical skill and creativity. Obviously, these assumptions are absurd.

Each facility has a certain inventory of equipment, which it owns and must amortize over the jobs that arrive. Some equipment performs certain jobs better or more efficiently than others. Different staffs, even more so, have unique experiences and varying skills. They may have creative solutions to your production needs, which could enhance the communication value of the production or reduce the time necessary to complete the project. For these reasons, a low bid may indicate a facility's strength. They may have produced projects similar to yours and have a creative solution ready at hand. The lowest-dollar bid, of course, should not be the exclusive basis for the decision to hire a company, but the numbers can hold clues to a distinctive competence, which may not be apparent.

WORKING WITH THE PRODUCTION COMPANY

There are some pitfalls in dealing with outside production companies that should be avoided. Do not base a graphics or animation budget for a new production on a project you have done before. Length of the finished product may be a crude guideline for estimating the price range for a graphics project, but the cost of computer graphics is based on the complexity of the piece, on the equipment employed to produce the sequence, and a number of other factors. A two-minute piece may cost a small fraction of a fifteen-second sequence, depending on the requirements of each. Get a new estimate from the production company each time you contemplate employing graphics; do not attempt to extrapolate past projects into a new budget.

Keep an open mind in dealing with the staff of the production facility. They can bring a wealth of experience to the project as well as a detailed knowledge of their own capabilities. A fresh perspective on your communications goal can sometimes result in a creative and rapid solution to the problem at hand. Elicit input from the staff of the commercial facility. That is, in part, what you can expect from them for paying their fee.

Coordinate communications with a facility through a single person. It is very confusing to have various interpretations of the producer's priorities. More important, the approval of graphics should be one individual's responsibility, and if the producer cannot tend to the review and approval of preliminary images, this responsibility must be delegated. Arguments over details of graphics, especially when they are stills from an animation, will deplete the budget in short order.

In a similar vein, do not hesitate to discuss concerns as they arise. Just as you expect your program to communicate, communicate with your vendor and align your expectations of the production along the way. Settle issues and concerns before they become problems that may cost money to resolve. Choose a vendor who is a good listener.

Develop a contract with the production house. Clarify the responsibilities of both parties in producing the video. Full knowledge of the scope of the project and the expectations of the finished product will go a long way toward keeping the production within schedule and budget and will minimize surprises and hard feelings along the way.

Most important, do not wait until the last minute to produce graphics or animation. Computer graphics take time to create, and any attempt to rush a project will compromise the quality of the product. The guidelines in Chapter 9 explain where and how early the artist should be included in the production process. Keep in mind, however, that when the graphics are coming from outside, even more time needs to be allocated. The independent production facility is usually involved in a number of projects simultaneously. Thus, it is not just a question of how long it takes to generate the appropriate images, but what the production company's schedule will allow.

If you have the flexibility, you may ask a production house for a *run of schedule*. This allows the facility to book you in a slot that works best within their schedule and allows you to finish the project on time. Similarly, many production houses will offer a discount from rate card for booking off-hours. Editing from 6 p.m. to 2 a.m. may be less expensive than prime time and may even make it easier for you to escape your own facility to devote time and attention to the project.

CONCLUSION

Commercial facilities are eager to be the exclusive subcontractor for a corporation. If a particular firm or number of free-lancers can provide everything a corporate producer needs when going outside the corporate facility, developing an exclusive relationship can go a long way toward facilitating jobs that go out of house. By using the recommendations in Chapter 9 to manage the relationship between you and your subcontractor, outside services can greatly enhance your product while also providing budgetary efficiency.

9 The Animator as Part of the Production Team

We now have some familiarity with the uses of computer graphics, the methods with which they are created, and the expense and time that a project may involve. Now we will turn to the available options for incorporating these techniques into a video production, and how the corporate producer can handle the introduction of computer graphics as a new production tool.

PLANNING IS THE KEY

Planning is a fundamental of media production. What must precede all else is the idea, the concept. Before video is selected, before computer graphics are chosen to display information, before the scripts and storyboards are written and drawn, the producer must understand the communications problem and formulate the most effective means of solving it. This may sound a bit trite, but it is a fundamental that needs to be dusted off when introducing computer graphics. Altering minute details of animated sequences can be exceptionally difficult and expensive, which reinforces the need for a thorough understanding of a production's goals up front. Having examined some of the ways in which computer graphics have been used and having some idea of their capabilities (especially within the constraints of the budget), we now turn to incorporating computer graphics into the production.

THE COMPUTER ARTIST AS SPECIALIST

We have made reference to the computer graphics artist as a specialist on the production team. Let us reinforce that reasoning as we begin our discussion of integrating computer graphics techniques into the producer's field of options.

It is not our intent to imply that a video producer or director is incapable of picking up the operation of a graphics system, nor do we believe that the artist or

119

computer operator by nature of the specialty has greater training or skill in visualizing the finished product or composition of an image on the screen than anyone else on the production team.

We do maintain, however, that the most effective means of utilizing a computer graphics system is by dedicating a specific person to perform that task as his or her principal function within the production department. The software, which creates graphics and animation, and the hardware, which runs it, are complex beasts. The more time one individual spends mastering the techniques, the more proficient the artist becomes. At the same time, the more thoroughly the artist understands the system, the more efficient her or she will be in the process of generating computer graphics.

INVOLVING THE COMPUTER ARTIST

Using this as a starting point, there are some principles that, when applied, assist in making a synergistic relationship between the computer artist and the other members of the production team. The guidelines for managing the artist or animator as part of the production team are the same whether computer graphics are created through an outside production facility or by internal staff and equipment.

Involve the artist early-on in the production process. This will contribute to the artist's understanding of the overall goals and strategy of the program. Contributions are most likely to be relevant and fit well with the rest of the content when the artist has been involved from the beginning. Just as important, knowing the expectations and goals of the producer within the context of the production will enable the artist to create computer graphics within deadline and budget. Waiting until post-production or later to bring in the graphics person will bog down the process. It will take time for the artist to come up to speed on the project. The artist will have less of an understanding of the goals of the program and how the producer intends to achieve them.

The artist may also have many creative contributions in the conceptual stages of the production. As we discussed earlier, the computer graphics specialist can often make suggestions as an art director. There may be some points that could be made more clearly with graphics, as well as some sections envisioned as graphics that might be better communicated with other means. Of course, specific imagery, which had not been conceived by the producer, can be suggested by the artist, and some of these images might generalize to other parts of the program in ways that had not occurred to anyone else on the staff. This arises not from any special talent, but from the artist's constant orientation to the communications medium. A savvy producer knows an online editor can make the difference between a good presentation and a great one. So, too, can the computer illustrator or animator.

This schedule of inclusion is a standard procedure at some production houses. One Pass Video in San Francisco, for example, routinely brings in the computer graphics artists for pre-production meetings with clients.

Include the editor in these meetings if you do not already. In Chapter 4 we discussed how graphics systems are often used effectively as an integral part of the post-production process. If graphics are enhanced with sophisticated post-production effects especially, the editor and artist must interact while planning and executing computer imagery for it to be effective and cost efficient.

When including these people in pre-production, allow some flexibility in your vision of the finished production. Let them contribute creatively, do not limit them to being technicians attempting to carry out your personal visual concepts. Not only will this permit and promote some innovative visuals, but you may end up saving significant production costs. It is not unusual for a skilled post-production team to take advantage of their system to generate images that approximate a desired effect, but are far less difficult to produce than the original concept. Take the opportunity to be flexible.

STORYBOARDS

Storyboards are critical. They do not necessarily have to be detailed renderings of the desired visuals; simple representations of the principal visual elements and major movements are sufficient. In animation, these may be (and often are) quick and dirty sketches of major design elements and motion paths. In fact, it may be a good idea not to develop detailed storyboards until the artist or an art director who will operate the computer or work with the computer operator is available. It is important to specify images that may be created efficiently by the computer system available to the producer. It is of little use to bring the artist detailed renderings of images that cannot possibly be produced within budget. A thorough knowledge of the system's capabilities and time requirements is necessary for adequate planning of the graphics.

Thoroughly check any information that will be presented graphically in the program. If any mistakes exist in the data you give to the artist for presentation and the errors are carried through to the rendered image, a virtual reconstruction of the image may be required. Any changes to finished images cost money in reworking time.

APPROVALS

Get necessary approvals as early on in the process as possible. Have key visuals generated and review them or take them to the parties responsible for approval. Review animated sequences in simplified form before rendering. Any changes to a finished piece of animation are, as a proportion of the graphics budget, fantastically expensive. Remember that animation is but a collection of still, indivisible pictures. Change any element of that image and the entire picture must be reconstructed. You effectively double the cost of rendering the sequence. If three-dimensional objects with specular surfaces are in the visual, if the work involves retouching single frames of video, or even if a two-dimensional image is manipulated in post-production and adjoins live video transitions, changes late in the production can blow the graphics budget right out of the water.

INTEGRATION

Work to facilitate communication between the computer artist and the client. Remember, you may be almost used to the spectacular effects possible with graphics systems, but many corporate clients are still fearful of committing themselves to some of these effects, in part because of the high cost and because the technology is still a mystery to many people. Lars Tragardh, a past president of Synthetic Video and an experienced hand at creating corporate and industrial computer graphics, points out that fear and mystery help sell a movie but do not encourage corporate clients to spend thousands of production dollars on a regular basis.

Remember that the computer graphics artist or animator, to be most effective as an asset to the producer, is not someone to be called in to contribute some pictures for inclusion in the production. The artist must be a member of the team, with the same responsibilities and opportunities for contributions to the creative aspects of the program as the other members. The artist should be included at the same time and in much the same way as the scriptwriter, the director of photography, and the editor. When the artist interacts with the rest of the team to formulate the tactic for communicating the content of the program, the producer will realize the full potential and make most efficient use of what computer graphics have to offer.

Glossary

Abekas: Video devices manufactured by Abekas Video Systems. Products include the A-42 digital still store, the A-52 digital video effects unit and the A-62 digital videodisc recorder, which can store up to 100 seconds of live video on a Winchester hard disk.

ADO (Ampex Digital Optics digital video effects unit): The ADO is capable of manipulation of the video raster in three-dimensional space.

AI (Artificial Intelligence): The ability of some specially programmed computers to reason, to learn and to simulate sensory capabilities.

airbrush: On a paint system, the technique of creating a soft line or field of color, similar to using a paint atomizer to create diffuse color on canvas, paper, metal or other surface.

Alias/1: Sophisticated 3-D computer animation system produced by Alias Research, Inc.

aliasing: The process by which a computer graphics system generates the illusion of a diagonal line on a video screen. Since the video image is comprised of square pixels, a true diagonal is impossible. Lines that deviate from vertical or horizontal are actually built from a series of squares laid corner to corner, creating the "alias" of a diagonal line. Unmodified diagonals from a graphics computer appear jagged, or appear to stair-step across the screen.

anti-aliasing: Software that attempts to overcome the jagged appearance of aliased lines. One of the simplest methods involves filling in the steps a diagonal line makes with a color that is the average of the line and background colors. In reality this decreases the resolution of the line, but it appears sharper by obscuring the "jaggies."

analog: A system of measuring or varying current by frequency or amplitude. In computers, it is the alternate to digital, which transmits information by the mere presence or absence of a signal. A digital thermometer has a readout of the temperature in numbers; the analog version would be the mercury thermometer.

animation: The rapid display of a sequence of images, which creates the illusion of motion.

Artronics: Producer of the VGS Video Graphics System and PGP Presentation Graphics Producer computer graphics systems.

Artronix: Producer of the Artronix Studio Computer computer graphics system.

Artstar IIID: Graphics computer produced by Colorgraphics Systems, Inc.

aspect ratio: Relationship of the length of an image to its height. Television's Aspect Ratio is 4:3.

Aurora Systems: Producer of the Aurora 220 computer paint/animation package.

AVA: Computer graphics system for television produced by Ampex. It was the first commercially available television graphics system; now in its third generation.

BEFLIX: Computer language developed in the 1960s to assist in producing computer animation; now obsolete.

bit: Short for binary digit; the smallest unit of digital information with a value of 0 or 1.

bit map: The translation of each pixel on the screen to a location in computer memory. It has the advantages of speed and flexibility, but requires a large amount of computer memory to be effective.

bit pad: See *digitizing pad*.

border: Line around part of a video screen defining the area of a key or wipe.

Bosch FGS-4000: 3-D computer animation system.

brush: Group of pixels, which is assigned to the tip of a stylus by a paint system. Assigning a color to the brush and passing it over a digitizing pad results in the color being copied onto each pixel location it touches, resulting in a line of color.

byte: A group of adjacent bits, which combine to form one unit of data.

CAD/CAM (Computer-Aided Design/Computer-Aided Manufacturing): The use of computers to assist in the design and drafting of products.

CAE (Computer-Aided Engineering): Similar to CAD, the use of computers in drafting and documenting anything from buildings to microchips.

caption: Subtitle identifying what is on a video screen or what the actors on screen are saying as an aid to the hearing impaired.

cathode ray tube: See *CRT*.

cel: A single clear sheet of acetate film onto which a cartoon character is painted. It is then laid over a background. The film once used was cellulose, from whence the name. In computer graphics, cel refers to a single frame of video.

cel animation: Traditional film animation technique of placing cels over a background on a special registration stand. The layered image is photographed on movie film, and new foreground cels are laid in place.

CGI: Computer-generated imagery.

chameleon: Low-cost paint system by Chyron.

Character Generator (CG): Specialized computer graphics system. In its simplest form, an electronic typewriter for placing words on a video screen for use as titles, subtitles, captions and so on.

chromakeying: See *key*.

Chyron: Producer of a family of character generators and graphics systems.

color cycling: Simple form of animation on a paint system; the illusion of motion is created by rotating colors around a single image. Also known as *color mapping*, *paint animation* and *look-up table animation*.

crawl: Text moving at a readable rate across the screen.

CRT (cathode ray tube): Vacuum tube with an electron gun at one end and a rectangular, flat screen at the opposite end, which is coated on the inside with electrosensitive phosphors. When struck by a passing beam of electrons, the phosphors glow. Directing the beam over a designated path or selectively turning the beam on or off as it passes specific locations on the coating can create images that are visible outside the tube.

Cubicomp: Producer of 3-D animation systems based on the IBM PC/AT.

cursor: A blinking character, often a square or dash, indicating the position on a monitor where the next computer input will be displayed.

cut and paste: Copying part of a video image and placing it somewhere else on the screen.

Cypher: Character generator with 3-D capabilities produced by Quantel.

data screen: In computer graphics systems utilizing more than one monitor, the monitor that displays information about the image and system status. Possible information displayed includes current disk name, current image name, space remaining on disk, current function selected.

dedicated system: A computer that has been customized for graphics production and cannot be utilized for other computer functions without cannibalizing or altering the system.

digital: Having to do with numbers. In computers, refers to the transmission and processing of information in binary form.

digital effects: Devices permitting manipulation of the video raster to accomplish movement, such a rotation, shrinking and panning. More sophisticated machines are capable of distorting an image as well as repositioning it to accomplish, for example, the appearance of a turning page.

digitizer: Device that feeds information in digital form to a computer. The device may be a video camera, an optical scanner, a digitizing pad or similar device.

digitizing pad: Tablet embedded with wire mesh whose intersections correspond to pixel locations on a screen. Inputs information to a computer by detecting the presence and location of an electronic stylus, mouse or puck on its surface. Also known as a *bit pad* or *digitizing tablet*.

disk operating system: Software that instructs a computer how to access and store information. Generally, the operating system is the first program run on the computer whenever the machine is turned on. Without it, most computers would lack necessary information for communicating with peripheral devices.

Display Processor Unit (DPU): Digital to Analog conversion device developed by Ivan Sutherland as part of Sketchpad, which allowed information output by the central processor of a computer to be displayed on a CRT or plotter.

dither: Smoothing a jagged line by adding points of color.

door swing: Rotation of a video image on a screen axis.

drop shadow: The effect of a shadow beneath an image, so that it appears to be suspended above the background.

DSS (Digital Scene Simulation): Largely attributed to Gary Demos and John Whitney, Jr. through their work at the now-disbanded Triple-I Productions. DSS is a method of digitally matting foreground action over a computer-constructed background entirely within a computer for subsequent output to a film recorder or videotape machine.

Dubner: Brand name of a family of computer graphics systems including the CBG-2 combination character generator, paint system and 3-D modeling computer.

DVE: Trademark of NEC for its Digital Video Effects unit, sometimes used as a generic name for digital effects.

DVST (Direct View Storage Tube): A device that interposes a wire mesh between the electron gun and phosphor coating in a CRT. It made possible the perpetuation of an image on the screen without requiring continuous refreshing by the computer. Although the images were low quality, it achieved rapid popularity due to its being inexpensive. The DVST became obsolete with the dawn of the inexpensive frame buffer.

embossing: The effect of raised patterns from a background.

Encore: Digital effects device produced by Quantel.

Euclidean geometry: Set of basic principles governing points, lines, planes and solids in one, two and three dimensions, developed by the Greek mathematician Euclid.

EXPLOR (Explicitly Provided 2-D Patterns, Local Neighborhood Operations and Randomness): An early computer graphics language developed in 1970 by Kenneth Knowlton.

extrusion: Creating a 3-D image from a 2-D pattern by adding depth to all sides of the image. Extruding a square would create a cube or rectangular box, depending on how far it was extruded, and extruding a circle would yield a cylinder.

Fairlight CVE: Computer Video Instrument, low-cost paint and effects machine.

fill: Command that defines a specific area to be a certain color.

firmware: Logic for performing certain functions within a computer, usually located on circuit boards, which can be connected to a central processor via wires or expansion slots.

floppy disk: Thin, flexible platter covered with magnetic media for storage of data.

font: Typeface alphabet.

fractal: Models constructed by the application of a branch of mathematics developed by Dr. Benoit Mandelbrot which utilizes intermediate dimensions to simulate irregular shapes found in nature.

frame: Single still video image. Frames on videotape are displayed at a rate of 30 per second.

frame buffer: Specialized computer memory, which temporarily stores a number of frames of video for manipulation or continuous display.

frame grab: Option on a computer graphics system that permits the capture of a video image from tape or camera for storage or manipulation in the computer.

frame store: Video device used to store video images for recall at some later time.

frisket: A form of electronic stencil that allows manipulation of a video image either within or outside the defined area.

front ending: The process of performing certain data communications functions in a smaller computer than the one that will execute the program, lessening the load on the executing computer.

generation: Each successive recording of a video sequence from one tape onto another. Each time a sequence is rerecorded from one tape onto another through processing or special effects equipment some noise is introduced. The goal of quality-minded producers is to distribute productions as close to first generation as possible.

GENESYS: Early graphics programming language.

Gouraud Shading: A method of shading polygonal meshes developed by Henri Gouraud. In order to make a surface of polygons appear smooth, Gouraud developed an algorithm to gradually change the shade of a polygon from one side to another so that it would blend smoothly with the surrounding polygons.

graftals: A rendering method developed by Alvy Ray Smith, which exhibits properties of fractals and particle systems, but which differs from both in that it does not incorporate randomness. It generates images by recursively dividing a form according to one basic rule a limited number of times. This process creates the illusion of detail.

graphics tablet: See *digitizing pad*.

grid: A pattern of horizontal and vertical lines placed at regular intervals. Initially incorporated in computer graphics as a positioning tool, it has found wide use as a design element in many video graphics.

growth algorithm: Method of image generation developed by Yoichiro Kawaguchi. In an attempt to simulate nature, this method does not construct images from discrete elements, but develops images from single points according to specified rules.

hard copy: Permanent and transportable computer output. Computer graphics hard copy is most often on paper or film transparencies.

hard disk: High volume, high density magnetic storage medium. On all but mainframe computers, the actual disks cannot be accessed physically by the operator.

Harry: Digital videodisc recorder by Quantel.

high resolution: A relative term used in video graphics to describe images that exceed the NTSC 525-line convention.

Illusion: Digital video effects device produced by Digital Services Corporation.

Images II: Paint and computer graphics system developed at the New York Institute of Technology (NYIT) and produced by Computer Graphics Labs.

IMI-600: Computer graphics device produced by Interactive Machines, Inc.

in-betweening: The process of interpolating images between key frames to relieve the computer artist from having to generate many similar illustrations in the process of animation. Sometimes called *tweening*.

interactive: System characteristic of immediate communication between the computer and the user.

interpolation: In *tweening*, the insertion of a specified number of images that metamorphose one image to another when animated.

jaggies: The broken or stair-step appearance of a diagonal line as a result of aliasing.

joystick: A vertical control stick often capable of deflection around a full 360°, which moves the cursor or image in the direction of deflection.

justify: Spacing type so that the text is positioned as desired between margins; lines of type are of a specified length.

kern: To create a letter that appears to extend beyond its side bearings.

key: A video effect in which one part of an image is replaced by another image or piece of an image. *Chromakeying* is the process of removing a specific color from an image, usually the blue or green, to cut a hole for the new image. *Multilevel keying* is also called *matting*, after the film process of creating composite visuals.

key frame: The beginning or ending frame of a simple motion sequence in animation.

Knox: Producer of low-cost, high-resolution character generators.

light pen: Electronic wand connected to a video monitor whose position over the image is denoted by a point of light on the screen. A button on the pen is usually used to select a function or position, which is highlighted on the screen.

logo: Distinctive design or typeface used to identify a company.

look-up table: Reference chart used to identify a limited number of addresses from a larger number of possibilities. In paint systems, the look-up table is used to form the palette of a small number of colors, which can be displayed on the screen, from a large number of colors, which the systems can distinguish.

mask: Area of the screen that is unaffected by a special effect or process.

matting: See *key*.

megabyte: One million bytes of primary or secondary computer storage capacity.

menu: A list of processing functions from which the operator may choose.

Meteostat: Computer colorization system used with satellite weather information to distinguish different meteorological conditions.

microcomputer: An arbitrary designation for a small computer, usually powered by microprocessors. Various parameters have been devised to differentiate micros from larger systems in terms of memory size and processing speed, but none are standard.

Mind-Set: Small character generator produced by JVC.

minicomputer: An arbitrary designation for a small computer, usually the size of a desk, whose operating capabilities fall between the microcomputer and the main-frame.

Mirage: Three-dimensional digital image manipulator produced by Quantel.

model animation: Film animation produced by photographing constructions of the character or object of the animation. The process usually involves matting and multiple layers of film.

motion control: Computerized movement of a camera used, for example, in the process of recording miniatures.

mouse: Small, box-like computer peripheral, which communicates position or motion information to a computer via a tracker ball located on its underside. Moving the mouse across the surface of a table results in rotation of the tracker ball, which then relays data about the degree and direction of this rotation to the computer.

multilevel keying: See *key*.

neon: Tubular, glowing line effect, which simulates neon lighting.

node: Vertex (corner) of a polygon.

opening: Introduction or beginning sequence of a program.

page: The amount of memory corresponding to one complete video image.

palette: The collection of colors used to compose a computer image. Also used to refer to the color menu displayed during the generation of an image.

paint animation: See *color cycling*.

paint system: Hardware and software package, which enables the operator to generate freehand images on a video screen by direct manipulation of a frame buffer. Paint systems usually consist of a computer processor, video screen, storage devices, frame buffer, digitizing pad and stylus or mouse.

Panther Graphics: Combination character generator and paint system produced by 3M.

particle system: Image generation algorithm developed by Alvy Ray Smith, which creates pictures by tracing the paths of computer-created points according to certain rules.

pass: One trip a video signal makes through processing or special effects equipment. Each pass is usually used to add a layer of effects so that a finished program may have many effects occurring at once. In traditional editing, each pass also implies one generation of tape. In some of the latest digital video effects equipment, however, multiple passes can be performed within the digital environment without the loss of quality inherent in multiple tape generations.

pen: See *stylus*.

PC: Personal computer; IBM trademark for their brand microcomputer.

PC-controlled system: Also called PC-based system. In computer graphics, a system constructed around a microcomputer, but which leaves the computer unchanged from the form in which it can be purchased for other purposes. Contrasts with a dedicated system, in which the computer is integral with the image processing equipment.

Phong shading: Computer procedure developed by Phong Bui-Tuong, which renders surface specular reflection.

phosphors: Electro-sensitive particles coated onto the inside of a video screen, which glow when struck by a passing electron beam.

Pixar: Firm marketing an animation and rendering package in a variety of specialized applications, including medical and geophysical analysis. Once a part of Lucasfilm's Industrial Light & Magic division.

pixel: Picture element, the basic building block of the video image.

pixellization: Breakup of the video image to achieve a tile or mosaic effect. Often incorrectly called "pixellation." A film effect of unnaturally fast motion usually employed for its comedic appeal.

polygon: Two-dimensional multi-sided geometric shapes; the enclosure created by a number of straight line segments. Polygons are frequently used to construct simulated surfaces or solids.

post-production: The stage of creating a video program occurring after studio or field production. Editing, special video effects and computer graphics are all part of the post-production phase.

posterization: Breakup of the video image into areas of intense and relatively pure color.

Prisma: Graphics and animation system produced by Digital Services Corporation.

prompt: A computer-generated message instructing the operator on the next anticipated input or procedure.

puck: Device similar in appearance and identical in function to the mouse.

Quantapaint: Graphics and paint system by Quanta.

Quantel Paintbox: Trademarked name for paint system by Quantel.

RAM disk: Temporary computer memory, created by the addition of additional internal memory, but addressed by the computer as if it were a disk drive. RAM disks are used for rapid access to internal memory when the ability of the microprocessor within the computer has been loaded with as much memory as it can address already.

raster: Collection of pixels, which forms a single video image.

raster graphics: Computer-generated images created by manipulation of pixels on the screen, as in a paint system.

raster-scan: Format of a raster graphic image. Contrasts with a *vector graphic*.

ray tracing: Computer rendering model developed by Phillip Mittleman and refined by Turner Whitted, which defines individual pixel values by looking back into the computer model of an image and reflecting off or refracting through the model until the illuminating light source or sources are encountered. Ray tracing is computationally demanding but renders the most realistic version of reflective, refractive and specular surfaces.

real time: In playback, the pace at which the final production or sequence will be viewed. In recording, the period of capturing images as they are created.

rendering: The process of converting a mathematical or geometric model into a screen image.

RGB: The three individual component color signals, which are composited to form a complete video signal. The signals are kept separate in component systems through post-production until they are ready for recording onto tape.

rotation in perspective: Foreshortening of a video image as it is turned by a digital effects device to give the illusion of depth.

rotoscope: To trace a filmed or taped frame and subsequently record the resulting drawing. In film, a frame of motion picture is projected and traced with pen, pencil or other drawing tool, and the sequence of drawings is shot on an animation stand. In video, single frames are redrawn in a paint system. The resulting sequence may be used independently or in addition to the live video.

rubberbanding: Stretching an image from a point or area. In drawing on a paint system, for example, a line or square can be pulled out of a single point. The line stretches from that fixed point to wherever the stylus is directed on the screen.

rubberstamp: The repeated copying of a design element on the screen.

Scanimate: Early computerized video effects device capable of a number of analog conversions of an image, such as spins and distortions.

scanned image: Video image captured by a computer graphics system for manipulation. See *frame grab*.

scene simulation: See *DSS*.

screen resident menu: Computer function controls, which appear on the same display as the image being manipulated. These functions can often be selected by the stylus on an unused portion of the digitizing pad.

scroll: See *crawl*.

segue: Transition between sequences.

serial sectioning: Creating a 3-D image by sequentially inputting parts or slices, which are then joined together by the computer.

single frame animation: Motion visuals created by producing images that get recorded onto consecutive frames of film or video.

smooth shading: Computerized rounding of the corners of polygonal structures and providing shadows and reflections.

stair step: The close-up appearance of an aliased line.

stencil: Paint system function of creating a mask to protect a certain part of an image from being changed by a subsequent process.

step frame: Deleting certain frames from a video sequence in order to create a strobe-like effect.

still store: A hard disk device for recording and storage of individual frames of video.

stochastic: Mathematical model possessing an element of probability or randomness.

stylus: Electronic pen used in conjunction with a digitizing tablet or touch screen to select functions or create images by identifying and transferring information to specific screen locations.

super: Slang for superimposition. See *key*.

surface model: Design of 3-D objects by manipulating a curved grid or polygonal mesh.

tablet: See *digitizing pad*.

terrain modeling: Computer simulation of surfaces found in nature, often accomplished by the use of fractals.

Texta: Character generator produced by Dubner.

texture mapping: The "wrapping" of a video image around a three-dimensional model, creating what appears to be a contoured surface.

titler: Character generator.

tomography: X-ray technique of photographing one plane of a body, creating a cross section or cutaway view.

touch screen: Computer terminal screen, which allows selection of a function by touching it with a finger or stylus.

tracker ball: Hand-operated control used for image placement or color control, or used as the control instrument on the underside of a puck or mouse.

translate: To move without rotation, often in a straight line.

tweening: See *in-betweening*.

user-friendly: Pertaining to a computer system that allows operators unfamiliar with the system to interact easily with it.

vector graphic: Image composed of lines and curves on a CRT whose electron beam traces over only those phosphors that will define the image. Contrasts with a *raster scan* image, in which the electron beam visits each pixel in turn, stimulating those that define the image.

Vidifont Graphics 5: Character generator and graphics system produced by Thompson CSF Broadcast.

virtual device interface: Communications software, which permits data exchange between incompatible computers.

Weather System: Computer graphics system connected to a satellite data system. Used to enhance meteorological information for television weather broadcasts. See also *Meteostat*.

Winchester Disk: High-capacity magnetic hard disk data storage device for microcomputers.

wire frame: Bent grid or polygonal mesh skeleton of a three-dimensional object.

x, y,-z axes: Parameters of three-dimensional space. Usually, the *x* axis is horizontal, the *y* axis vertical, and the *z* axis runs parallel to the line of sight of the viewer into the screen.

zoom: Visual movement along the *z* axis into or out of the screen.

Bibliography

BOOKS

Brush, Judith M. and Douglas P. *Private Television Communications: Into The Eighties*. Berkley Heights, NJ: International Television Association, 1981.

Goodman, Cynthia. *Digital Visions, Computers and Art*. NY: Harry N. Abrams, Inc., 1987.

Jankel, Annabel, and Morton, Rocky. *Creative Computer Graphics*. Cambridge, MA: Cambridge University Press, 1984.

Magnenat-Thalmann, Nadia, and Thalmann, Daniel. *Computer Animation - Theory and Practice*. Tokyo: Springer-Verlag, 1985.

Rivlin, Robert. *The Algorithmic Image*. Redmond, CA: Microsoft Press, 1986.

PERIODICALS

Address, Eric R., and Muderick, Michael. ''Creating New Visual Demands.'' *Video Systems,* February 1986.

Allen, Joe. ''Belles Lettres.'' *Millimeter,* May 1985.

Amato, Mia. ''Another World.'' *Millimeter,* April 1985.

Amato, Mia. ''Glossary of Computer Graphics Tools, Terms and Techniques.'' *Millimeter,* April 1986.

Anderson, Beth. ''Charting the Course of Computer Graphics.'' *Videography,* February 1986.

137

Anderson, Beth. "Electronic Graphics and Animation on Display." *Videography,* June 1986.

Anderson, Beth. "What's What in Computer Imagery." *Videography,* November 1986.

"Animation/Special Effects Facilities." *Millimeter,* December 1986.

Chudnoff, Richard. "Production Leaders Assess CGI Marketplace." *Computer Pictures,* September/October 1985.

Chunovic, Louis. "Paint Box or Paintbrush?" *Millimeter,* Feburary 1985.

"Conversation with Robert Abel." *Videography,* February 1986.

Cook, Tom. "Creating Cost-Effective Computer Graphics." *Video Systems,* February 1986.

Davis, Mills. "The Ideal Electronic Studio." *Computer Pictures,* May/June 1985.

Dawson, Denise. "The Invisible Paintbox and Other Magic Tricks." *Videography,* February 1986.

Doyle, Claire. "Graphics on a Budget." *Television Broadcast,* November 1985.

Dunn, John. "Painting by Pixels." *Audio-Visual Communications,* February 1985.

Ebersole, Phil. "Are Gnomes Running Kodak?" *Democrat & Chronicle,* December 20, 1987.

Emett, Arielle. "Universal Studios Computer Graphics." *Computer Graphics World,* February 1986.

Foreman, Joel. "It's Only a Mirage." *Video Systems,* February 1986.

Gowin, Steven. "San Francisco's New Perspective on Video Graphics." *Videography,* August 1986.

Grimes, Bruce. "Managing Outside Resources." *Video Systems,* June 1987.

Heller, Neil. "Animation Propagation." *Video Systems,* July 1986.

Herzfeld, Jill. "KDS Turns Ideas Into Powerful Animated Effects." *On Location,* November 1985.

Hurn, Bruce. "Paint Animation." *VideoPro,* July/August 1985.

Hutzel, Ingeborg. "Digital Paint and Animation Systems Overview." *Millimeter,* April 1985.

Johns, Alison. "Agencies in Wonderland." *Millimeter,* February 1985.

Jones, Charles. "Computer Animation in Industrial Videos." *On Location,* February 1986.

"Just Wild About Harry." *Video Graphics and Effects,* Spring 1986.

Keckan, Maria. "Contracting Outside Services." *Video Systems,* June 1986.

Lu, Cary. "Micros Get Graphic." *High Technology,* March 1986.

MacNichol, Gregory. "Getting Graphics Onto Video." *Computer Graphics World,* May 1986.

MacNichol, Gregory. "A PC-Based Animation Primer." *Computer Graphics World,* November 1986.

Meeks, Thomas D. "Moving the Message When Money is Tight." *Educational and Industrial Television,* September 1985.

Meyers, Paul, and Knorr, Eric. "The AT Animates the Airwaves." *PC World,* February 1986.

Murray, Frank. "The Charlex Way." *Videography,* February 1986.

"Postproduction Facilities Survey." *Videography,* August 1985.

Rich, Nancy. "Affordable Animation." *Audio-Visual Communications,* January 1986.

Rivlin, Michael. "Shopping Guide to Graphics Production Companies." *Computer Pictures,* July/August 1986.

Rivlin, Robert. "New Frontiers for Animation." *VideoPro,* November 1985.

Robertson, Barbara. "Micro-Based Video Development Surges." *Computer Graphics World,* January 1986.

Robertson, Barbara. "Video Control for the IBM PC." *Computer Graphics World,* November 1985.

Schloss, Alfie. "Graphics Grow in the Aisles." *Television Broadcast,* March 1986.

Schubin, Mark. "Cheap Thrills." *Videography,* June 1985.

"Shopping Guide to 3D Modeling and Animation Systems." *Video Graphics and Effects,* Spring 1986.

St. Lawrence, Jim. "Low-Tech Special Effects." *Videography,* February 1986.

Stokes, Jim. "Adding Special Effects to Corporate Videos." *Video Systems,* September 1987.

"Teleproduction Facilities (various regions)." *Millimeter,* December 1986.

Tragardh, Lars. "Animated Conversation." *Videography,* September 1987.

Uibel, George. "So You Wanna Get Into Computer Graphics?" *Video Systems,* July 1986.

Van Deusen, Richard. "What's Your Technical IQ?" *Video Manager,* May 1987.

Weinstock, Neal. "From Maybridge to Microprocessors — Computers Bring Animation up to Date." *Millimeter,* February 1985.

Wershing, Stephen, and Weinberger, Tanya. "Corporate Communications Come to Life with Animation." *Educational and Industrial Television,* March 1986.

Wolf, B.A. "Graphic Demonstrations Provide Quick Communications." *Educational and Industrial Television,* March 1986.

Index

About the Authors

Stephen Wershing is a marketing and financial consultant. He was Marketing Director of Telesis Productions, Inc. in Rochester, NY and served as manager of the Media Center at Nazareth College. Stephen Wershing began his work in video while studying marketing and management at the Rochester Institute of Technology where he served as Promotions Director and later as General Manager of the television station.

Paul Singer is a freelancer in the public relations field. He has experience in all aspects of video production and was news cameraman for WOKR-TV in Rochester, NY. He has been a freelance aerial photographer, videographer, producer and director for Telesis Productions, Inc.